AutoUni – Schriftenreihe

Volume 179

Reihe herausgegeben von

Volkswagen Aktiengesellschaft, Volkswagen Group Academy, Volkswagen Aktiengesellschaft, Wolfsburg, Deutschland

Daniel Fuchs

Exploration of the Influence of Additive Manufacturing Design Potentials on Vehicle Conception

Springer Vieweg

Daniel Fuchs
Technische Universität Braunschweig
Braunschweig, Germany

Any results, opinions and conclusions expressed in the AutoUni Schriftenreihe are solely those of the author(s).

ISSN 1867-3635 ISSN 2512-1154 (electronic)
AutoUni – Schriftenreihe
ISBN 978-3-658-49676-0 ISBN 978-3-658-49677-7 (eBook)
https://doi.org/10.1007/978-3-658-49677-7

This publication was supported by the NiedersachsenOPEN publication fund, enabled by zukunft.niedersachsen.

This Springer Vieweg imprint is published by the registered company Springer Fachmedien Wiesbaden GmbH, part of Springer Nature.
The registered company address is: Abraham-Lincoln-Str. 46, 65189 Wiesbaden, Germany

If disposing of this product, please recycle the paper.

Preliminary Remarks

Additional References

Sec. 3.2 of this thesis is based on the author's previous work:

Fuchs, D. "Potenziale additiver Fertigungsverfahren für die Gesamt-fahrzeugkonzeption". Master Thesis, Prof.Dr.-Ing T. Vietor, Institut für Konstruktionstechnik, Technische Universität Braunschweig, July 2017

Translation Challenges

Translation challenges: A crucial part of the literature of the three research fields relevant to this work (Design Methodology, Vehicle Conception, and Design for Additive Manufacturing) is written in German. On the one hand, this leads to incomprehensible references for non-German speakers. On the other hand, translation challenges from German to English arise considering the following main terms in this thesis:

Konstruktion: The meaning of the English terms *Design, Engineering Design, Construction,* and *Styling* can be ambiguously related and are all used to translate aspects of the German term *"Konstruktion"* in the literature. Therefore, besides the term definitions in this work, the term differentiation in the literature is not clear. For this work, the term *Engineering Design* is chosen as a most suited translation of the German term *"Konstruktion"* and is thus mostly used.

Konstrukteur: Similarly the German term *Konstrukteur* is also translated ambiguously in the literature. A few examples are: *Designer, Engineering Designer, Design Engineer, Construction Engineer,* and *Stylist.* For the context of this work, the term *Design Engineer* is chosen to be most suited to translate the German term *Konstrukteur* and is therefore used in most cases.

Abstract

This thesis analyzes the potential impact of additive manufacturing-related design freedom on vehicle conception. In this context, the term *Additive Manufacturing* (AM) incorporates the technology field often referred to as *3D-Printing*. The motivation for this work is derived from alternative design solutions presented in state-of-the-art AM use cases, in which additional design freedom is leveraged to achieve significant product-and process-related improvements. This thesis aims to systematically explore the impact of AM-related design freedom on vehicle conception, leveraging novel design potentials from the part to the complete vehicle level.

Based on the fundamentals of the research fields *Design Methodology* (DM), *Vehicle Conception* (VC), and *Design for Additive Manufacturing* (DfAM) state-of-the-art approaches to AM-based vehicle design are analyzed, and key application categories are identified. Engineering design challenges derived from those categories, e.g., *mass individualization* and *functional integration*, are addressed by elaborating tangible automotive use cases. DfAM application principles are then leveraged from the part to the complete vehicle level, focusing on lightweight potentials and alternative approaches to vehicle body design. Necessary adaptions to development approaches, computer-aided design workflow, and tools are derived and suggested.

The customer-oriented DfAM-focused application concepts and engineering design approaches developed in this work conclude how additional AM-based design freedom might impact upcoming vehicle conception.

List of Publications

Fuchs, D.; Kuschmitz, S.; Kühlke, K.; Vietor, T.; Lachmayer, R.; Rettschlag, K.; Kaierle, S. (Eds.). „Identifikation von Zielkonflikten bei der Anwendung von Potenzialen additiver Fertigungsverfahren". Konstruktion für die Additive Fertigung 2019, Springer Berlin Heidelberg, 2020, 223–244.

Tschorn, J. A.; Fuchs, D.; Vietor, T. „Potential impact of additive manufacturing and topology optimization inspired lightweight design on vehicle track performance". International Journal on Interactive Design and Manufacturing (IJIDeM), 2021, 15, 499–508.

Fuchs, D.; Bartz, R.; Kuschmitz, S.; Vietor, T. „Necessary Advances in Computer-Aided Design to Leverage on Additive Manufacturing Design Freedom"International Journal on Interactive Design and Manufacturing, 2022.

Kuschmitz, S.; Fuchs, D.; Vietor, T.; Lachmayer, R.; Bode, B.; Kaierle, S. (Eds.). „Customer Benefit Oriented Approach on the Application of Additive Manufacturing Potentials Based on Product Property Trade-Off's ". Innovative Product Development by Additive Manufacturing 2021, Springer International Publishing, 2023, 305–317.

Patent pending DE 10 2019 211 824 A1: Kumke, M.; Hartmann, I.; Fuchs, D.; Demircan, F.; Wehmann, Y. „Modulares System und Verfahren zum individuellen Konfigurieren eines Kraftfahrzeugs". Deutsches Patent-und Markenamt, 2021.

Patent pending DE 10 2020 116 463.7: Fuchs, D.; Hartmann, I.; Kumke, M. „Verfahren zur Erstellung von Realisierungskonzepten für Bauteilgeometrien unter Berücksichtigung verschiedener physikalischer Energieflüsse". Deutsches Patent-und Markenamt, 2021 [FHK20]

Patent pending DE 10 2020 208 323 A1: Kühlke, K.; Fuchs, D. "Haltevorrichtung für ein Fahrzeug, Verfahren zur Herstellung einer Haltevorrichtung sowie Fahrzeug". Deutsches Patent-und Markenamt, 2021

Contents

Nomenclature

Acronyms

3DP	3D Printing
AM	Additive Manufacturing
BAAM	Big Area Additive Manufacturing
CAD	Computer-aided Design
CAM	Computer-aided Manufacturing
CAx	Computer-Aided tools in general
CDO	Conflicting Development Objectives
CFD	Computational Fluid Dynamics
CNC	Computerized Numerical Control
DDM	Direct Digital Manufacturing
DfAM	Design for Additive Manufacturing
DfM	Design for Manufacturing
DfMA	Design for Manufacturing and Assembly
DfX	Design for X
DM	Design Methodology
DP	Design Process
DT	Design Theory
DTM	Design Theory and Methodology
ESM	Electronic Supplementary Material
FDM	Fused Deposition Modeling
FEA	Finite Element Analysis
FEM	Finite Element Method

GM	Generative Manufacturing
GMK	Geometric Modeling Kernel
GSD	Generative Shape Design
HMI	Human Machine Interface
ICT	Information and Communication Technologies
IMA	Imagine and Shape
LOM	Laminated Object Manufacturing
LR	Linear Regression
MC	Mass Customization
MDO	Multidisciplinary Design Optimization
MI	Mass Individualization
MPO	Multiphysics Design Opitmization
NVH	Noise, vibration, and harshness
OEM	Original Entity Manufacturer
ORNL	Oak Ridge National Laboratory
PD	Product Development
PDP	Product Development Process
PLC	Product Life Cycle
PSL	Principle Stress Line
QFD	Quality Function Deployment
RM	Rapid Manufacturing
RP	Rapid Prototyping
RT	Rapid Tooling
SE	Systems Engineering
SLA	Stereolithography
SLS	Selective Laser Sintering
SOP	Start of Production
STL	Standard Tessellation Language
SUV	Sport Utility Vehicle
TRIZ	Theory of Inventive Problem Solving
TUS	Technical Up-Scaling
UI	User-Interface
UX	User Experience
VC	Vehicle Conception
VDI	Verein Deutscher Ingenieure
WLTP	Worldwide harmonized Light vehicles Test Procedure

Symbols

α	Gradient angle [°]
$\ddot{\phi}_i$	Angular acceleration [rad/s^2]
Δc_D	Difference in down force coefficient [-]
Δm	Mass difference as weight-saving potential [kg]
η_u	Drive train transmission efficiency
ρ	Air density [kg/m^3]
A	Front cross-sectional area [m^2]
a	Acceleration [m/s^2]
B_e	Fuel consumption [g/m]
b_e	Specific fuel consumption [g/kWh]
B_r	Force of brake resistances [N]
c_d	Drag coefficient (e.g. 0,25–0,4)
d_{CoG}	Distance from to the center of gravity [m]
F_{Acc}	Force of the acceleration resistance [N]
F_{AD}	Force of aerodynamic drag [N]
F_{RG}	Force of slope resistance [N]
F_{RR}	Force of rolling resistance [N]
f_r	Rolling resistance coefficient (e.g. 0,017 at 150 km/h)
g	Gravitational acceleration [m/s^2]
i	Summation index
I_z	Yaw inertia [kg·m^2]
J	Mass moment of inertia [kg·m^2]
m	Vehicle total mass [kg]
n	Sum of summation iterations

P	Vehicle power [kW]
r_i	Radius [m]
t	Time [s]
v	Velocity [m/s]

List of Figures

Introduction

1

This thesis concerns the potential impact of additive manufacturing-related design freedom on conventional vehicle conception. In this context, the term *Additive Manufacturing* (AM) incorporates the technology field often referred to as *3D-Printing* (3DP). The manufacturing technologies in this field set themselves apart from conventional manufacturing processes, e.g., through their functional principle of producing objects by selectively adding instead of, e.g., subtracting, molding, or forming raw materials. Over the last decade, the field of research *Design for Additive Manufacturing* analyzed the design benefits derived from this basic functional manufacturing principle. The product- or process-related potentials assigned to AM technologies are considered disruptive by some and also caught the interest of the automotive engineers subsequent to, e.g., the aerospace and medical sectors.

The term *Vehicle Conception* (VC) refers to the early phases of the vehicle development process in the automotive industry. The main characteristics and features of the vehicle are defined at this stage. Here, it is crucial to consider the technological and economic feasibility of a vehicle concept, preventing significant issues at the subsequent stages of the development process [PS21, 1292 ff.]. Therefore, automotive engineers rely on a profound knowledge of the capabilities and limitations of available manufacturing processes to develop feasible vehicle concepts. Considering that automotive engineering is undergoing a paradigm shift from part- to function-oriented development approaches, this aspect is particularly relevant due to the increasing complexity of today's and upcoming vehicles [CKZ14; Con+14].

Based on these two aspects mentioned above, the introduction to this work is structured into four subsections. First, the general motivation and resulting

---

Supplementary Information The online version contains supplementary material available at https://doi.org/10.1007/978-3-658-49677-7_1.

  1

D. Fuchs, *Exploration of the Influence of Additive Manufacturing Design Potentials on Vehicle Conception*, AutoUni – Schriftenreihe 179,
https://doi.org/10.1007/978-3-658-49677-7_1

hypotheses are formulated (Sect.1.1). Next, previous work on related topics is outlined (Sect.1.2). The main objective and related research questions are explained afterward in Sect.1.3. The introduction chapter concludes with this thesis's approach and structural overview summarized in Sect.1.4.

1.1   Motivation and Hypotheses

The general motivation of this work is derived from alternative design solutions presented in the state-of-the-art in DfAM. Those solutions often proclaim disruptive product- and process-related improvements due to the utilization of additional design freedom on the part-level of the product. To leverage those design potentials further, the motivation of this thesis is to explore AM-related design potentials with a focus on developing complex products consisting of multiple parts, such as a vehicle, from a holistic perspective. This motivation implies at least three different aspects that are formulated as hypotheses below:

1. **Enhanced design freedom originating from additive manufacturing technologies has the potential to improve future vehicle concepts substantially.** Here, it is hypothesized that if design engineers have to consider different or fewer manufacturing restrictions, the solution space they derive their ideas from expands, optimizing traditional designs and leading to product or process innovations.

2. **Enhanced design freedom not only results in alternative design solutions but also demands alternative design strategies, approaches, and methods to enable design engineers to holistically leverage the multitude of AM design potentials known in the DfAM context.** This hypothesis implies that if additional design freedom is applied systematically, the influence of DfAM on VC might exceed the optimization of the vehicle's individual parts or components. Figure 1.1 illustrates different hypothesized relations of the potential impact of DfAM on a vehicle's performance depending on the DfAM level of application qualitatively. Those relations shown in Fig. 1.1 represent different potential progressions of leveraged effects achieved by applying AM design potentials throughout the product.

3. **With the increased manufacturing flexibility assigned to AM technologies, feasible variant-related production complexity increases, resulting in novel business opportunities for the automotive industry.** This hypothesis considers the strong dependency of today's vehicle production on tool-based manufacturing processes. Thus, the number of feasible design variants can be severely

limited, hindering additional value chains toward product customization, personalization, or individualization. Considering this manufacturing flexibility and the potentially increased shape, functional, material, and variant complexity, decreased production complexity and increased development complexity are expected in the context of additively optimized products. Figure 1.2 qualitatively illustrates an exemplary differentiation in the distribution of development complexity along the product development process comparing design for conventional manufacturing (tool-based) and DfAM.

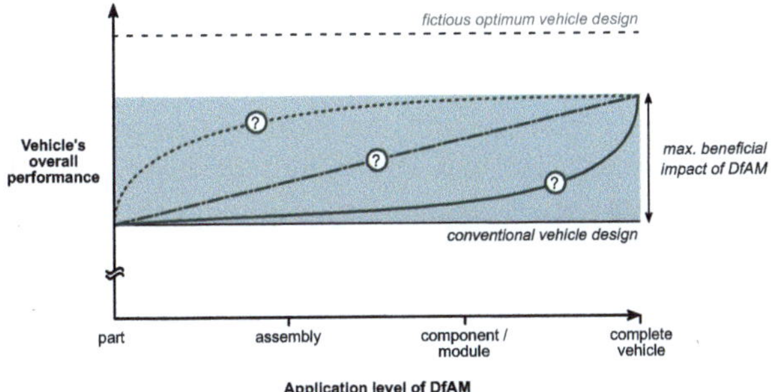

Fig. 1.1 Potential beneficial impact of DfAM on a vehicle's performance depending on its application level displayed qualitatively

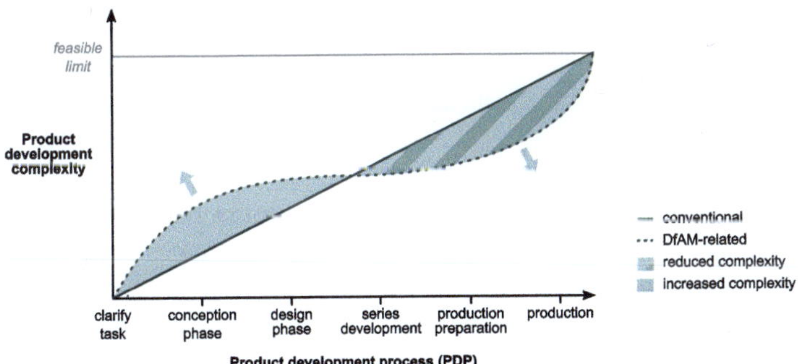

Fig. 1.2 Complexity shift from production planning phase into conception phase of product development process

1.2   Previous Work

The capabilities and limitations of AM technologies are profoundly analyzed and described in the literature and often referred to as design potentials and restrictions in this context [BCP12, p. 235; Geb16, 464 ff. GKT16, 139 ff. Gib+21, 559 ff.]. Several showcases and first applications in series production already demonstrate those manifold potentials, though often focusing only on single parts or small assemblies of a product.

Also, in the automotive industry, the first steps are being made to increase awareness of design potentials and utilize AM's manufacturing flexibility. Besides rapid prototyping, several product-related applications take advantage of AM design potentials on a complete vehicle level. Those concepts can be categorized as show cars, low-requirement vehicles, individualized vehicles, high-performance, and generative vehicle concepts. These use cases known to the state-of-the-art mostly take advantage of one-sided AM potentials, leaving room for improvement in the application of holistic DfAM methods suggested in the literature [BLR13, p. 13; Kum18, pp. 45, 58, 82]. Use cases with higher production focus and well-advanced technological readiness are limited to the part level in which optimized additively manufactured parts substitute conventionally designed ones. At this point, it is questionable if a part-substitution-approach exhaustively takes advantage of AM potentials for the vehicle.

Moreover, process-related innovations in rapid prototyping, jigs and fixtures made instantaneously at the assembly line, and new business opportunities in the product individualization context are being accessed. Latter meets the identified growth in the demand for diversification in our society during the last two decades [HSPI05, p. 2; Kor+15, p. 70]. According to this demand, flexible manufacturing technologies, e.g., processes not based on additional molds and tools, empower mass individualization. In the automotive context, first efforts are made to develop additively manufactured individualized products and introduce them to the market [LRZ06, p. 250; Tuc+08; Wag+16; Kab+17]. Simultaneously, the variety of vehicle configuration options for the end customer and thus his involvement in the product characteristics are drastically reduced by variant management methods to gain control over in-house complexity and increase companies efficiency [LRZ06, p. 8]. Therefore, AM's individualization and decentralization potentials conflict with automotive mass production strategies to reduce variant-related corporate complexity.

In this contradiction, additional work is necessary to meet the demand for development methods toward complexity management to utilize mass individualization potentials related to AM.

„*Complexity for free*" is a term commonly referred to in the AM context [HCD03, p. 26; Gib+21, p. 7]. It mostly emphasizes the often-claimed, non-existent correlation between geometric complexity and manufacturing costs for AM parts. This relation is contrary to conventional processes based on, e.g., tooling in which complex geometries require sophisticated and expensive tooling or machining to be feasible [HCD03; Fer+18; Gib+21, pp. 7, 15]. As noticeable in the paragraphs above, AM potentials are diverse and potentially induce additional complexity factors to be considered when implementing product development. Further, the claimed decreased production complexity, e.g., due to no need for tools and molds, can shift to the product's development phase, as taking advantage of AM-related design freedom can lead to increased shape, hierarchical, functional, and material complexity of a part. These potentially increased development efforts related to AM parts motivate the term „*complexity is not for free*" and require new engineering approaches, methods, and tools.

DfAM guidelines and methods suggested by literature already support design engineers to cope with increased design complexity [Con+14; KWV16; Tho+16; LL20]. Unfortunately, there are only a few commercial-ready computer-aided tools. Considering the limitations of conventional CAD and CAE tools, design engineers cannot fully leverage AM design potentials with reasonable resources. Therefore, the manufacturing capabilities of AM processes tend to exceed current development departments' abilities regarding engineering design, functional product structure, simulation, and optimization. Even with above-average development abilities, companies' bearable resources, e.g., the time available for design iterations, must be increased to develop highly optimized components of this kind. Therefore, most AM series applications can be observed in industries with high development budgets, e.g., aerospace, healthcare, and respective cost structures.

Figure 1.3 summarizes the white spots in literature mentioned above. It resembles the context of this work.

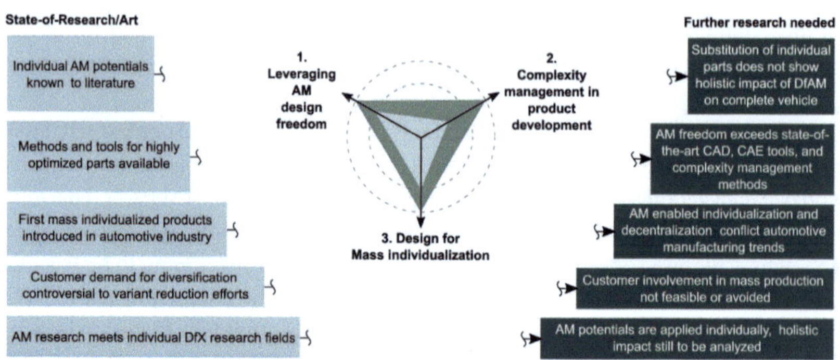

Fig. 1.3 Overview of state-of-research findings and gaps in the field of research

1.3 Objective and Research Questions

The main objective of this work is to systematically explore the potential impact of AM-related design freedom on vehicle conception, considering challenges within the intersection between three research fields: Design Methodology (DM), Design for Additive Manufacturing (DfAM), and Vehicle Conception (VC). Moreover, this thesis aims to contribute answers to the following three distinct research questions:

1. **How to apply DfAM potentials oriented on customer-relevant product performance in the vehicle conception context?** This question implies that it is crucial to consider AM-related design freedom during product development, yet additional customer benefits should be targeted in the first place. Customer benefits are often not the main focus in use cases presented in state-of-the-art AM applications. On the one hand, AM use cases are often designed with conventional manufacturing restrictions in mind. On the other hand, exaggerated complex geometries are additively manufactured and displayed as showcases. Thus, additional value for the product's performance or cost-efficiency is not leveraged holistically.

2. **Which vehicle property improvements can be expected by a holistic and systematic consideration of AM-related design potentials during vehicle conception?** The second research question focuses on identifying and analyzing vehicle properties that might benefit from additional AM-inspired design freedom when applying DfAM methods correctly, enabling this way a comprehensive assessment of the importance of the technology for automotive applications on part and complete vehicle level.

3. **How do contemporary engineering approaches, methods, and tools need to evolve to enable design engineers to utilize AM design potentials for vehicle conception holistically?** During vehicle conception, profound knowledge regarding design feasibility is critical to realizing vehicle concepts with an appropriate compromise between optimizing the product's performance and the effort spent during development and production. With extended design freedom due to shifted boundaries of what is possible to manufacture, contemporary design approaches, methods, and computer-aided design and engineering tools in vehicle conception are challenged by increased shape, functional, hierarchical, and material complexity enabled by AM. This research question addresses necessary advances during product development to enable design engineers to exploit AM potentials holistically.

1.4   Approach and Structure

The approach illustrated in Fig. 1.4 has been developed to address the main objectives and the research questions of this thesis mentioned above. The chosen approach consists of five main steps, either addressing the complete vehicle level or focusing on specific automotive applications. This separation is considered necessary to cover conception aspects regarding the vehicle in its entirety as a complex product while enabling detailed analysis of specific engineering challenges, relations, and principles in tangible use cases on part and assembly levels. This approach also resembles the structure of this thesis.

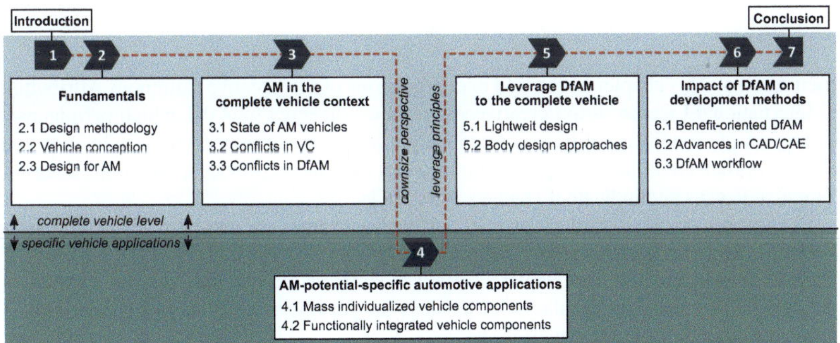

Fig. 1.4 Two-level thesis approach to address research questions

First, the fundamentals of the relevant research fields DM, VC, and DfAM are summarized in Chap. 2. In Chap. 3 the key areas for product-benefit oriented application of AM potentials on complete vehicle level are identified based on conflicting objectives in VC. Transitioning from complete vehicle level to specific applications, Chap. 4 focuses on solving challenges regarding the engineering design of mass-individualized vehicle components (Sect. 4.1) and functional integration (Sect. 4.2) on tangible use cases on part or assembly level. In Chap. 5 the leverage effects of AM potentials are analyzed in the context of DfAM-enabled lightweight design (Sect. 5.1) and alternative approaches to vehicle body design (Sect. 5.2), transitioning back to the complete vehicle level of the approach. The impact of the findings from Chap. 3 to 5 on engineering design methods and tools are then consolidated in Chap. 6. In Chap. 7, major findings are summarized and discussed, concluding this work with recommendations for future research. Additional content supplementing the abovementioned chapters is attached in the electronic supplementary material (ESM) appendices chapters available online.

Fundamentals 2

To elaborate on the research questions introduced in Sect. 1.3 this chapter covers the basics of the three main contextual research topics of this thesis. First, fundamentals regarding the field of research *Design Methodology* (DM) set the frame in the product development context (Sect. 2.1). Second, insights into *Vehicle Conception* (VC) outline the relevant scope of the automotive subject (Sect. 2.2). Third, *Design for Additive Manufacturing* (DfAM) outlines the most important findings of contemporary research to be brought in correlation with automotive conceptual design (Sect. 2.3). To put those three research topics into perspective, Fig. 2.1 illustrates DM in intersection with VC and DfAM as fundamentals of the following elaborations, still pointing out that each of these three individual research fields is part of a greater subject.

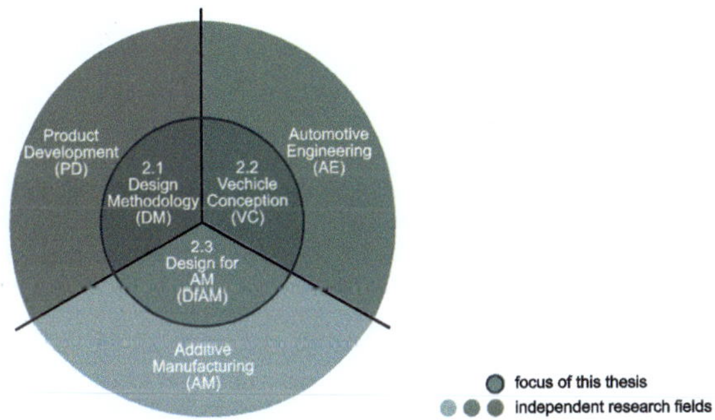

Fig. 2.1 Overview of the three fields of research in focus of this thesis and their relations to major subjects

2.1 Design Methodology in the Industrial Product Development Context

The development phase is significant for the successful realization of a product in an industrial environment and occupies the early stage of the *product development process* (PDP). The PDP describes all necessary steps between *product planning* and *start of production* (SOP), where the *product life cycle* (PLC) starts (Fig. 2.2). The development phase begins as soon as a product is planned and its main requirements are defined. In the development phase, not only the product's composition itself is engineered, but also its entire life cycle influenced. Therefore, the development phase is decisively responsible for the product's holistic performance in use. With this in mind *Product development* (PD) as a term can be defined in different ways:

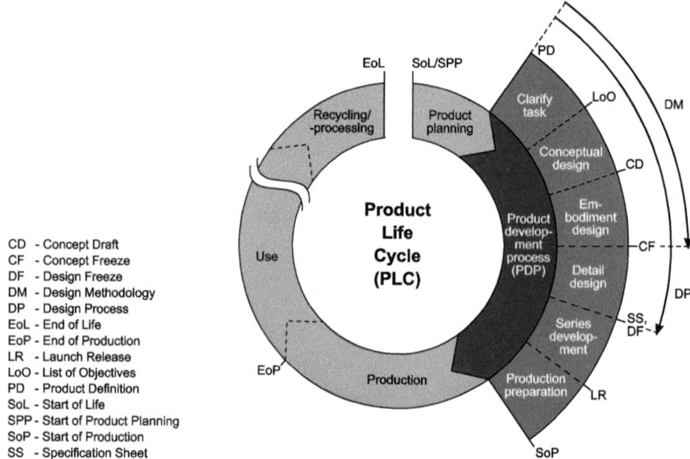

Fig. 2.2 Focus of design methodology in the product development process (PDP) and product life cycle (PLC) context based on [PS16, p. 1258; VDI93, p. 9; VDI19b, p. 13; VDI04, p. 18; And+12, p. 19; FG13, p. 23; Kum18, p. 39; AHC15, p. 24; EM13, p. 163]

Definition 1: *Product development* (user perspective) is the creation and launch of products with new or different functions or properties, or both, which offer new or added value to the customer. *Product development* (company perspective) is the use of exploration, design, manufacture, and marketing/sales to launch new, product-based businesses utilizing the company's resources. [AHC15, p. 21].

Even though Fig. 2.2 illustrates the major phases the PD and PLC in reality often several products are being produced and developed in parallel. Particularly in the automotive industry product generations and product upgrades follow each other closely often in 2–3 years turns. This leads to early phase development departments handing over projects to series development phase and starting over with the design phase for the upcoming product upgrade or generation even though the previous project has not even reached the start of production yet (Fig. 2.3).

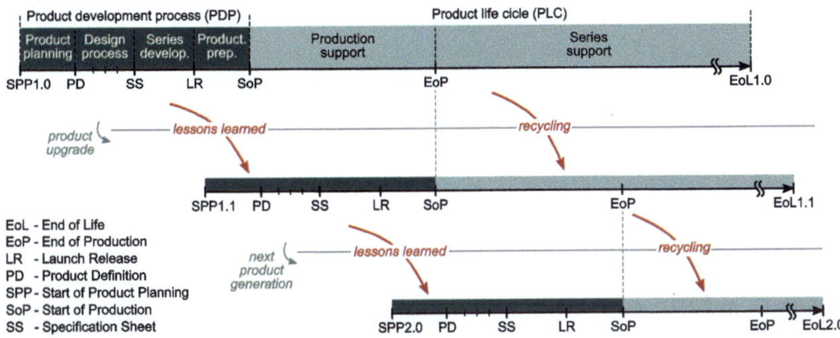

Fig. 2.3 Product development process and product life cycle considering product upgrades and product generations in the automotive context [PS16, p. 1258; VDI93, p. 9; AHC15, 227 ff.]

Besides the actual feasibility of a product, at least three main factors need to be fulfilled to ensure its success on the market (Fig. 2.4 *left*). To address customer needs and demands with its product line up and product characteristics is hereby one of the main objectives of a company. To do so, several aspects, such as reaching the market at the right time and with the right price offer, need to be considered during PD to achieve customer satisfaction (Fig. 2.4 *right*).

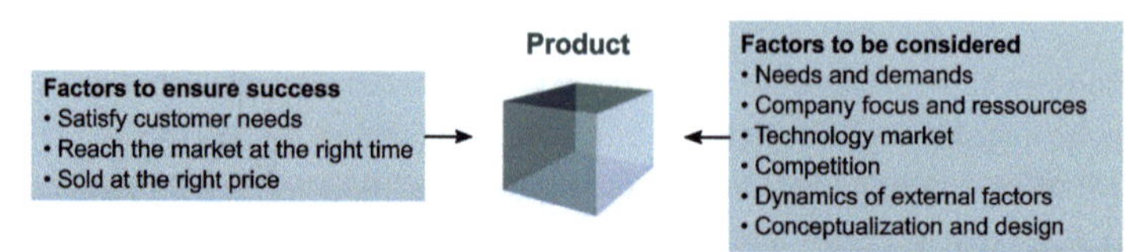

Fig. 2.4 Relevant factors for successful products based on [Pah+07, p. 134; AHC15, p. 21]

According to ANDREASEN ET AL. (2015) industrial PD includes all activities from ideation to launch, takes into account the whole life cycle, and incorporates the material cycle from raw supply to recovery and reuse. PD describes a totality independent of the type of product being developed, e.g., new, incremental, or platform-based products [AHC15, p. 21]. It is structured in consecutive methodological steps resulting in the PDP (Fig. 2.2). Note that the PDP might differ between application type and corporate environment. For further information consult [Neh14, B.1] as NEHUIS (2014) gathers and structures a variety of differing PDPs.

One of the main steps of the PDP is the engineering design itself. This step is also subject to a procedure, the so-called *design process* (DP). According to VDI (1993) the DP includes at least four main phases: *Planning and Task Clarification, Conceptual Design, Embodiment Design and Detail Design* [VDI93, p. 8; Nae18, p. 2]. It is, therefore, a crucial step of the PDP and has a substantial impact on the PLC. Due to the extended role of the DP in the PDP, both processes are often not differentiated from each other (compare [VDI93, p. 8]). Even though the boundaries between PDP and DP are blurred due to the iterative nature of product development in general, Fig. 2.2 illustrates that certain steps, e.g., performance testing, quality control, and others belong rather to the scope of the PDP than to the hands-on DP itself.

DM covers several steps of the DP and is one focus of the elaborations in this thesis (Fig. 2.2). Exhaustive theoretical elaboration of the topic of DM is accessible in the literature. See e.g. [Pah+07; NL20; EM13; Ehr+14; FG13; Lin09]. Before looking further into the definition of the term *design methodology* itself, the context of both words is put into perspective independently as follows:

> **Definition 2:** The partial term *design* is in this case mostly related to the context of *Engineering Design*. The main task of design engineers is to apply their scientific and engineering knowledge to the solution of technical problems and additionally the optimization of those solutions within the requirements and constraints set by material, technological, economic, legal, environmental and human-related considerations. While solving those problems a designer executes at least four main activities: conceptualizing, embodying, detailing and computing [Pah+07, pp. 1–5]

Even though engineering design is a crucial phase of the product development process, it remains just one of several segments of a development department in a company [EM13, p. 8]. Here, a systematic and structured working method is crucial considering the technical and environmental complexity of engineering tasks.

> **Definition 3:** *Methodology* is a body of methods, rules and postulates, a procedure or a set of procedures employed by a discipline [Web21]. It stands for the doctrine of scientific engineering methods. A *method* is a target oriented, planned and prescriptive approach based on sequential actions and rules in order to achieve a certain goal or solve a certain problem [Lin09, p. 57; EM13, p. 748].

Together both terms describe the field of research DM, closely related to the doctrine of *design theory* (DT). Due to the contextual proximity of both terms, they are often referred to in a collective manner *design theory and methodology* (DTM). Even though it is challenging to define and differentiate DTM clearly, TOMIYAMA ET AL. (2009) describe the classic view on both term segments as follows. On the one hand, DT is about how to model and understand design. On the other hand, DM is about how to design [Tom+09, p. 544]. Latter is one focus of this thesis. To be more precise, DM itself is defined differently in the literature. PAHL ET AL. (2007) chose a definition as follows:

> **Definition 4:** *Design methodology* (DM) is a concrete course of action for the design of technical systems that derives its knowledge from design science, cognitive psychology and from practical experience in different domains [Pah+07, p. 9]. In more detail design methodology is about three main points [Nae18, p. 51]:
> 1. Breaking down complex designs into manageable components without losing sight of the overall purpose.
> 2. Enable a consequential and purposeful working process.
> 3. Find the right tools in form of methods to achieve optimal results efficiently.

One main advantage of profound knowledge and experience in DM is that it facilitates design engineers to master design challenges successfully [NL20, p. 12]. To enable an overview of additional arguments for the necessity of DM and its related challenges Table 2.1 is introduced. Even though DM induces several benefits to the product development process, Table 2.1 also points out that the actual application of design methods in an industrial environment can be demanding for several reasons. Thus the challenge to develop universally valid methods that are still applicable to specific use-cases can be mentioned as one of many. Nevertheless, due to the efforts accomplished in DM and its broader sense in DTM, it is now possible to teach engineering design [EM13, 9 f.].

However, development methods should be chosen carefully as well as specifically to the company's environment and adapted if necessary [Ehr+14]. In the following subsections insights into exemplary basics of DM (Sect. 2.1.1) as well as into design and optimization tools (Sect. 2.1.2) are given. As one particular and subsequent relevant category of PD, an overview of specifics towards product development for *mass customization* is summarized in Sect. 2.1.3.

Table 2.1 Benefits and challenges in the context of *Design Methodology*

Benefits	Challenges
EHRLENSPIEL AND MEERKAMM (2013)	EHRLENSPIEL AND MEERKAMM (2013)
• Grasp complex problems/tasks despite limitations of human brain	• Application/integration into industrial context
• Enables specialization and teamwork due to explosion of knowledge	• Display effectiveness of methods
• Cope with increasing complexity of systems	• General validity versus applicability to specific use case
• Deal with increasing time pressure	• Often focused on new designs (*10% of the case in practice*) rather than re-designs and variant-design
• Counters explosion of diversity and variety of products, opinions,etc.	• Consideration of all possible solutions versus acceptable results in short time in practice.
PAHL ET AL. (2007)	• Enough flexibility regarding its scenarios of application
• Fosters and guide the abilities of designers	• Integration in CAD tools
• Promotes creativity	• Human centering
• Enables objective result evaluation	PAHL ET AL. (2007)
• Render design theory comprehensible	• Compatibility with concepts, methods and findings of other disciplines
• Enables the subject to be taught	• Not rely on finding solutions by chance
• Provides reusable solutions	• Facilitate application of known solutions to related tasks
• Divides the work between designers and computers meaningfully	• Be compatible to electronic data processing
LINDEMANN (2016)	• Be easily taught and learned
• Support product development, conception and design	• Reflect findings of cognitive psychology and modern management science
• Error prevention	• Ease planning and management of teamwork
• Support for planning of development phases	• Provide guidance of product development team leaders
• Enhance communication of involved parties	

2.1.1   Basics of Design Methodology

According to the DM literature, design tasks in industrial context can be classified in three main categories: *original, adaptive* and *variant designs* [EM13, p. 12; Pah+07, p. 4]. Even though in most industrial cases design variations (70%) and design adaptations (15%) occur, the majority of methods suggested in the field of research design methodology address original design tasks [EM13, p. 12; Tom+09, p. 562]. Nevertheless, elements of those methodologies are also suitable for design adaptions or variants.

To clarify the scope of DM TOMIYAMA ET AL. (2009) categorize the DTM field roughly into four main criteria along two axes; one is *abstract* vs. *concrete* and the other is *general* vs. *specific*. As shown in Table 2.2, design methods are often derived from specific applications and concrete in their description (e.g., how to design a vehicle drive train). DM consists of factual knowledge and methods that have been generalized to apply to several use cases. Math-based methods are often obtained through the abstraction of specific design methods in order to implement them algorithmically, e.g., in computation. Generalized and abstract design knowledge is assigned to DT. [Tom+09, p. 544]

The observations elaborated in this thesis include approaches contributing to specific and concrete design methods and general DM. Several standards and norms describe state-of-the-art fundamentals of DM and can be consulted for implementation purposes. Moreover, the contemporary literature elaborates on not yet established approaches, guidelines, and methods for engineering design. Therefore, state-of-the-art and state-of-research are introduced in the following.

Table 2.2 Categorization of design theory and methodology (DTM) with focus on design methodology (DM) and methods based on [Tom+09, p. 544]

	General	*Specific*
Abstract	Design theory	Math-based methods
Concrete	**Design methodology**	**Design methods**

Standards, Norms and State-of-the-Art Development Processes

Standards and norms set and document guidelines to give orientation and unify undesirable diversity in engineering. Those standards are not necessarily mandatory but serve as guidance on how engineering tasks can be solved according to the state-of-the-art of each specific field of expertise. The *Verein Deutscher Ingenieure* (VDI) developed several guidelines addressing the context of PDP, DM and DP [VDI93; VDI19b; VDI19c; VDI97; VDI82; VDI04]. Those guidelines are often referred to in the literature. Even though the classic development procedure from VDI Guideline 2221 [VDI93] valid from 1993 has been updated to VDI Guideline 2221 Part 1 and 2221 Part 2 [VDI19b; VDI19c] in 2019, the main approach suggested to DP in PD remains in its core the same. It consists of four main phases and distinguishes at least seven stages among those phases (Fig. 2.2, 2.5). It becomes clear that the design process incorporates more than just the modeling procedure of parts that represent the product when assembled. *Phase I – Task clarification* is subsequent to product planning and sets of the DP. The following phases start abstract (*Phase II – Conceptual Design*) and become increasingly concrete (*Phase III – Embodiment Design*) with each additional step concluded, finalizing the design with the elaborations of the details (*Phase IV – Detail Design*). Distinctive results are assigned to each task of the procedure that, in sum, result in the product documentation. However, the final solution needs to meet the specification sheet requirements to proceed with the product realization. After the DP, minor changes to the design might still be necessary due to new findings obtained during product testing. Nonetheless, at some point the *Design Freeze* (DF) need to be set in order to avoid countless iterations (Fig. 2.2).

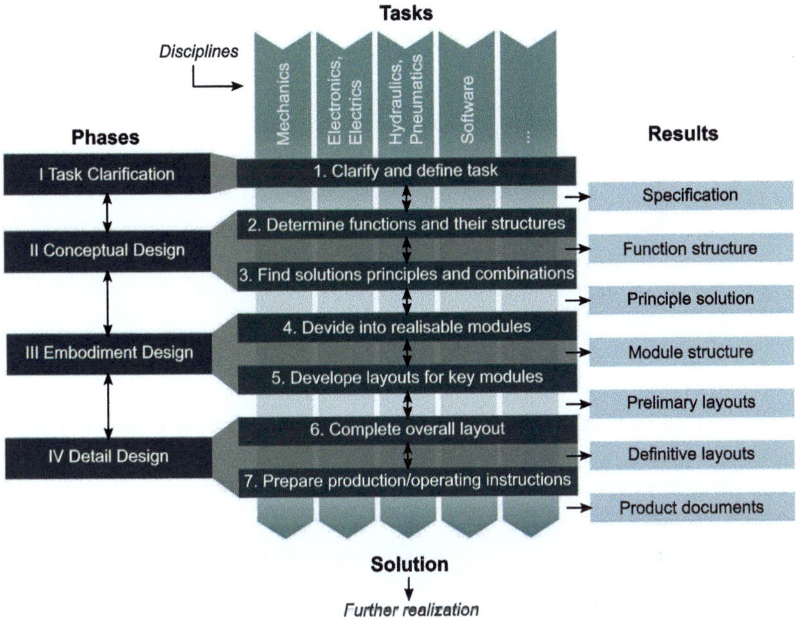

Fig. 2.5 General engineering design process in product development based on [VDI87; VDI19b; VDI04] (compare also [Pah+07, p. 130])

Since complex products often consist of several technical disciplines that require differing experts/departments to work together, the DP introduced in Fig. 2.5 should be considered as a simplified and generalized version of the DP in a corporate environment. Therefore, in the guideline VDI-2223 [VDI04] a complemented version of the DP is introduced. Here exemplary technical disciplines such as *Mechanics*, *Electronics*, *Pneumatics* and *Software* are displayed as parallel work streams throughout the main tasks of the DP from VDI-2221 [VDI87] emphasizing the necessity of multidisciplinary cooperation and teamwork in the DP (Fig. 2.5).

The increasing functional complexity of products due to the rapid development of digital product functionalities demand adapted approaches in PDP. Engineers increasingly perceive technical products as complex systems consisting of functional architectures. The term *systems engineering* (SE) is therefore gaining importance in automotive development departments.

> **Definition 5a:** A *system* is defined as a set of interrelated components working together toward a common objective [Kos+20].

> **Definition 5b:** A *system* is a model of an object (a real or conceived product or activity) based on a certain viewpoint, which defines the elements of the system and their relations. A system carries structure, i.e. the elements and their relations (arrangement, architecture) and behavior, i.e. the system's response to a stimulus depending on stimuli, structure, and state [AHC15].

Overall, SE is seen as an integral part of project management in which it plans and guides engineering efforts [Kos+20]. KOSSIAKOFF ET AL. (2020) define SE as follows and simultaneously differentiate it to traditional engineering disciplines.

> **Definition 6:** The function of *systems engineering* is to guide the engineering of complex systems. It differs from mechanical, electrical and other traditional engineering disciplines in several ways [Kos+20]:
> 1. It is focused on the system as a whole, it emphasizes its total operation.
> 2. It is concerned with costumer needs and operational environment.
> 3. It leads system conceptual design.
> 4. It bridges traditional engineering disciplines and gaps between specialties

The strength of SE is that basic features can be combined with technical theories and models to create more powerful models [AHC15]. In this context, the *V-Model* can be mentioned as one established development approach in SE. It has its origins in the development of mechatronical systems, and it aims to establish a cross-domain solution concept that describes the main physical and logical operating characteristics of the future product. For this purpose, the overall function of a system is broken down into main sub-functions. Suitable operation principles or solution elements are assigned to these sub-functions. Furthermore, in this model, the performance of a function is tested explicitly in the context of the system [VDI21b].

Figure 2.6 illustrates how, during the design of a product in the automotive context, domain-specific design tasks are derived from the overall system and solved independently (Fig. 2.6 *left side*). During their integration back into the overall system frequent assurance of properties is carried out (Fig. 2.6 *right side*) [VDI21b]. Consequently, short control loops are established during PD reducing negative consequences further down the line due to lessons learned on the way [FG13].

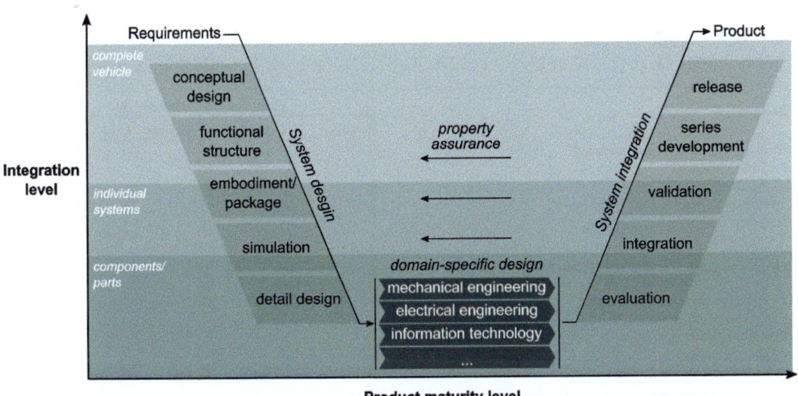

Fig. 2.6 Integration of key steps of the automotive PDP into the V-model based on [VDI21b, p. 29; Web09, p. 12]

When analyzing the development of standards and norms considered in the PDP, a trend becomes clear that emphasizes the importance of a function-oriented approach in the DP. Further, it becomes crucial to comprehend complex technical products as systems consisting of several levels, again comprising functions of different domains to cope with increased product complexity.

Besides the procedures introduced in the technical guidelines above, illustrating *what* needs to be done during the PDP, numerous methods describe *how* to approach these development steps. Collections of those specific methods can be found in state-of-the-art guidelines. Table 2.3 shows an overview of exemplary methods applicable for each phase of the DP. Five different categories are being distinguished. The first category *analysis and goal-setting methods* contains approaches especially applicable for the early phase of the DP, setting the proper foundation by providing guiding requirements for the upcoming design phases. The second category *development of solution ideas* is necessary throughout the process and often applied in iterative fashion. These methods are primarily used to open up the solution space, containing ideas suited for the early concept phase and others rather useful during the detailing phase of the DP. Besides the development of appropriate solutions, the third category consists of techniques and procedures for *evaluation and decision-making* support design engineers to reduce the number of solution ideas down to the most expedient ones and to choose the optimal overall solution effectively. When

Table 2.3 Exemplary method collection containing suited methods for each phase of the DP based on [VDI93, p. 33; Kum18, p. 31; NL20, p. 27]. (● = well applicable; ○ = applicable)

Method	Phases of the Design Process			
	I. Planning	*II. Conception*	*III. Embodiment*	*IV. Detailing*
Analysis and goal-setting methods				
Definition of goals	●	○		
Competition analysis	●	○		
Specification sheet (list of requirements)	●	○	○	○
Function structure (function means tree)		●	○	○
...				
Methods for development of solution ideas				
Creativity Methods (brain storming, 6-3-5 method, etc.)	●	○		
Morphology (morphological matrix)		●	○	○
Design rules and guidelines (design principle catalogs)		○	○	●
Heuristics (systematic combination of basic principles with applications)		●	○	
...				
Evaluation procedures and decision techniques				
Dual comparison	○	●	○	
Use-value-analysis	○	●	●	●
Relevance or decision tree		●	○	○
...				
Integrated methods (action models)				
Systems engineering (V-model)	●	●	○	○
Design methodology	●	●	●	●
Computer-aided design (CAD) and manufacture (CAM)		○	●	●
...				
Cost calculation and economic assessment procedures				
crucial in the PDP, yet not particular focus point in the DP				
...				

targeting to cope with interdisciplinary design problems systematically, the fourth category *integrated methods and action models* should be considered. Procedures for *cost calculation and economic assessment*, the fifth and last category, should also be taken into account throughout the PDP. Nevertheless, the procedures of the fifth category are not the focus of the DP in the context of this thesis.

State-of-Research in Design Methodology

Potential future research topics of DM expressed in the literature correlate with the general challenges of DM gathered in Table 2.1. Already TOMIYAMA ET AL. (2009) suggest to set research focus on product complexity and multi-disciplinarity. The consideration of increasingly complex requirements and multiple stakeholders with different cultural and educational backgrounds is still a challenge today. Additionally, the management of complex PDP needs improvement, e.g., by integrating differing engineering domains even closer to each other. Linked to the globalization trend, additional collaboration techniques between engineers are necessary. Here, the integration of advanced *information and communication technologies* (ICT) for computer-oriented design methodologies with a focus on virtual engineering is expected. [Tom+09]

Further developments of traditional methods [Orl17] and the combination of those methods with simulation and optimization approaches can be observed in recent research efforts [DM+20; XI18]. As an example, the *theory of inventive problem solving* (TRIZ) can be mentioned. The realized importance of innovative design by the industry led to gained popularity of, e.g., TRIZ as a method to enhance innovation capabilities [Tom+09]. Even invented by the scientist GENRIKH SAULOWITSCH ALTSHULLER at the end of the 1940s in Russia [Orl17], TRIZ is continuously being further developed, adapted, and combined up to recent days.

However, besides the modernization and combination of traditional methods, new frameworks are also being proposed to address the contemporary challenges of DM. Exemplary, the framework for the development of complex products introduced by XUE AND IMANIYAN (2018) can be mentioned [XI18]. Based on a computer-aided parameter optimization approach, it considers modeling, simulation, and optimization aspects during the concept and detailing phase (taking non-geometric and geometric descriptions into account).

According to the variety of requirements on today's products (e.g., cost-efficiency, lightweight, manufacturability, sustainability, modularity, and others), research on DM also explores rather specific engineering design challenges. In this context, the term *Design for X* (DfX) has been established further in the last decade (where X could stand for engineering design objectives such as assembly, manufacturability, safety, reliability, customization, among others). Here, specific design guidelines, approaches, methods, standards, norms, and rules are being developed to support design engineers during the DP, focusing on particular product properties or phases of the PLC [RS12, p. 446; FG13, p. 476; EM13, p. 357; Mit+14, pp. 17; Kum18, p. 33]. Definitions of the term DfX differ from each other in the literature, yet they can be summarized as follows:

> **Definition 7:** *Design for X* (DfX) summarizes all methods, strategies and tools that enable design engineers to consider several aspects and influences, with relevance to the PLC, simultaneously during PD [Bau03, p. 1]. Further, DfX has two meanings. The variable X can either stand for specific product properties (e.g. costs, quality, lightweight) or for a certain phase of the PLC (e.g. assembly, recycling) [AM97].

Figure 2.7 gives an overview of differing exemplary manifestations of DfX and suggests their structuring in relation to the PLC. Each combination (e.g., Design for Assembly, Design to Cost, Design for Recycling, and others) refers to a set of design guidelines and methods to support design engineers to purpose design a product according to an individual requirement or PLC phase in focus during the DP. Regarding the number of manifestations, it becomes clear that DfX is com-

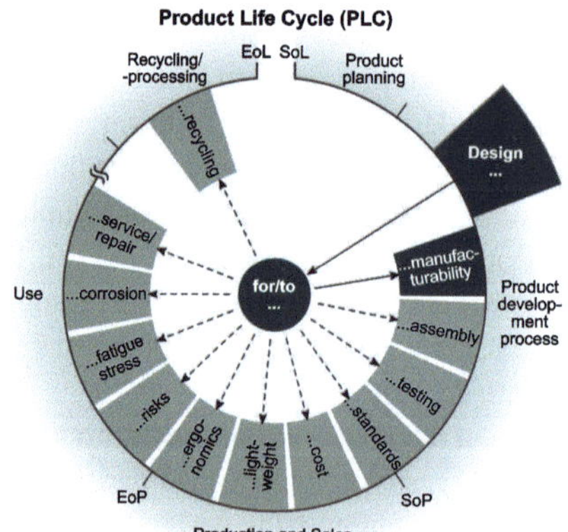

Fig. 2.7 Exemplary Design for X (DfX) overview structured according to the PLC based on [RS12, pp. 446–448; Bau03, p. 1; Mit+14, pp. 17; Kum18, p. 33; AM97, p. 7]

plex and challenging to consider holistically during PD. Moreover, this topic is constantly expanding according to trends in research and industry (e.g., Design for Mass Customization 2.1.3, Design for Additive Manufacturing 2.3.2).

EHRLENSPIEL AND MEERKAMM (2013) give a tabular overview of the relevant literature related to different DfX manifestations [EM13, p. 357]. Furthermore, MITAL ET AL. (2014) point out that most DfX tools fail to make a clear distinction as to when and how they should be used but merely provide a list of recommended design rules with little direction on their use [Mit+14, p. 280]. Additionally, they emphasize that the simultaneous optimization of many differing design goals (DfX) is the next step in research [Mit+14, p. 270].

In conclusion, several challenges are being dealt with in the state-of-research on DM. Traditional methods are being adapted and combined with modern approaches to cope with higher product complexity. Additionally, usability and applicability of methods are being increased by focusing on the development of computer-aided applications (digital catalogs with design guidelines and rules, wiki-based knowledge databases, implementation of design methods into CAD tools, and others). In general, the simultaneous optimization of differing design goals throughout the DP still occupies researchers' minds searching for holistically optimal design approaches and solutions.

2.1.2   Computer-Aided Design and Optimization Tools

Besides theoretical approaches and methods of DM, design engineers rely on *computer-aided tools* (CAx) to solve complex design tasks. In the following paragraphs relevant fundamentals of CAx tools with focus on *computer-aided design* (CAD) are summarized. For this purpose, insights into CAD systems' categorization, composition, and actual design workflow are given. The evolution of design tools illustrated in Fig. 2.8 emphasizes the gained importance of computer-aided design and development tools in the past. Starting as simple 2D geometry models and calculation algorithms (in the 1950s) computer-based design software is nowadays defining entire products in virtual space, including fully digitized back-end development processes. For an extended version of the CAx history consult, e.g., [Vaj+18, pp. 8–13].

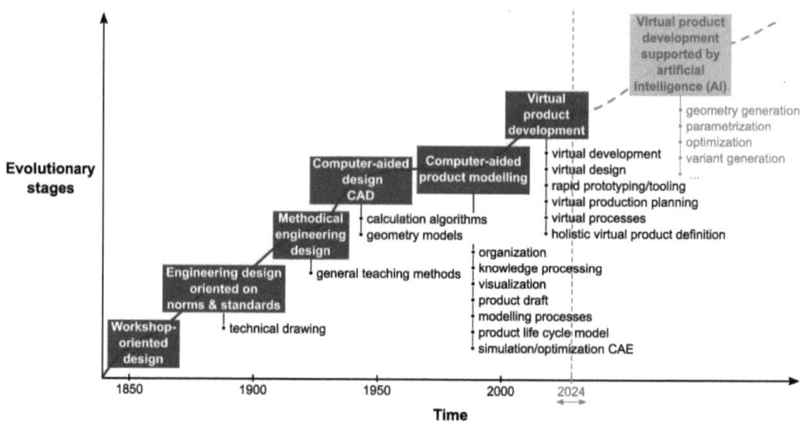

Fig. 2.8 Qualitative overview of milestones in the history of development and design tools free translated and based on [SK97; FG13, p. 9; Vaj+18, pp. 8–13]

Definition 8: *CAx-Systems* (CAx) is the summarizing term for all computer aided systems in a company. The variable „x" is a placeholder for multiple acronyms specifying the area of application (e.g. „D" for „Design", „E" for "„Engineering", or „M" for "„Manufacturing"). As a standalone term CAx means *computer-aided everything* and is from here on understood as the systematic application and consequent development of computer-aided methods, approaches and tools in the product development context (translated from [Vaj+18, p. 1])

Categorization of Computer-Aided Engineering Tools

Starting with the output of 2D engineering drawings in the 1960s, nowadays one of the main objectives of CAD systems is to define a physical product regarding its 3D geometric shape (1980s), correct assembly, structure, and system components. This way, CAD is crucial for engineers to figure out how to produce (*Computer Aided Manufacturing* CAM) a complete product and how it might perform once in service (CAE). [KBF05]

Nowadays, computational support can be provided continuously throughout the PDP, e.g., through CAD, CAE, CAM, CIM, PDM, and PLM systems reducing design and production effort [Pah+07]. Excessive literature is already available covering this topic [Vaj+18, pp. 15; Pah+07, p. 62; FG13, p. 416; RS12, p. 890]. Nevertheless, the categorization of those systems, acronyms, and terms is important to gain a brief overview of design and optimization tools in use by design engineers in corporate environments today.

VAJNA ET AL. (2018) summarize and explain several systems and their acronyms in the context of CAx. Table 2.4 lists an extraction of those systems, pointing out the wide scope of CAx solutions in the engineering context. It is helpful to match

Table 2.4 Explanation of common acronyms in the CAx context translated from [Vaj+18, p. 15] and complemented with [Wer+04]

Acronym	Term	Explanatory Comments
CACD	Computer-Aided Conceptual Design	
CAD, CAGD	Computer-Aided Design, Computer-Aided Geometric Design	*additional yet out-dated meanings: Computer-Aided Draughting/Drafting/Drawing*
CAAD, CAS	Computer-Aided Aesthetic Design, Computer-Aided Styling	
CAID	Computer-Aided Industrial Design	
CAE (FEA/FEM,CFD)	Computer-Aided Engineering The most relevant sub-categories are: - FEA/FEM: Finite Element-Analysis/-Method - CFD: Computational Fluid Dynamics - Tools for Dynamics Simulation - CAO: Computer-Aided Optimization	*mostly used in the simulation context, occasionally used for Computer-Aided Electronics*
CASE	Computer-Aided Software Engineering	
CAT	Computer-Aided Testing	*occasionally Computer-Aided Tolerancing*
DMU	Digital Mock-Up	*used to build up digital prototypes*
VR	Virtual Reality	*fundamental technology for DMU*
KBE	Knwoledge-Based Engineering	*implementation of knowledge processing systems in PD*
RP, RT, RPT	Rapid Prototyping, Rapid Tooling	*closely related to 3DP/AM technologies*
CAM (CA(P)P, CNC, MES)	Computer-Aided Manufacturing The most relevant sub-categories are: - CA(P)P: Computer-Aided (Process) Planning - CNC: Computer Numeric Controlled - MES: Manufacturing Execution System	*focuses on manufacturing, assembly and testing processes by...* *planning,...* *programming machines and ...* *executing orders*
CAPL	Computer-Aided Plant Layout	
PDM	Product Data Management	*management of product data during development phase, closely related to ERP in production*
CIM	Computer-Integrated Manufacturing	
CAQ	Computer-Aided Quality Assurance	
ERP	Enterprise Resource Planning	*management of all operating equipment and resources of a company*
PPS/PPM	Production Planning and Management	
PLM	Product Life-cycle Management Systems	

those systems to the respective phases of the PDP to clarify on which of them design engineers rely during the development of a product. WERNER DANKWORT ET AL. (2004) categorize a majority of those systems into different steps of a general PDP chain [Wer+04, p. 1440]. Additionally VAJNA ET AL. (2018) emphasize the importance of major systems by locating them accordingly to their use in the PDP as well as illustrating the necessity of the specific CAx system concerning the overall product complexity [Vaj+18, p. 19]. Fig. 2.9 is based on these two preliminary works and also emphasizes those CAx systems related to the context of this thesis.

Figure 2.9 emphasizes that CAID, CAD, CAE, and CAP tools are crucial during PD. Further, it becomes clear that even though CAD tools could be considered the principal tool for design engineers in the past, CAE systems are gaining importance due to steadily increasing product complexity. Nowadays, CAD and CAE systems are highly interlinked with each other. Therefore, the implementation of both formerly separated environments into modern holistic engineering software ecosystems is one major ongoing development trend in this area. It is becoming increasingly challenging to distinguish the exact functional scope of CAD and CAE tools today. On the one hand, CAD software gets enhanced with simulation- and process-planning-based functions, and on the other hand, CAE software gets extended with basic design functions. This correlation reveals the limitations of CAD. Today's geometry focus of CAD is of little help when reasoning about, e.g., functions, interactions, and activities [AHC15, p. 195]. Among other things, several

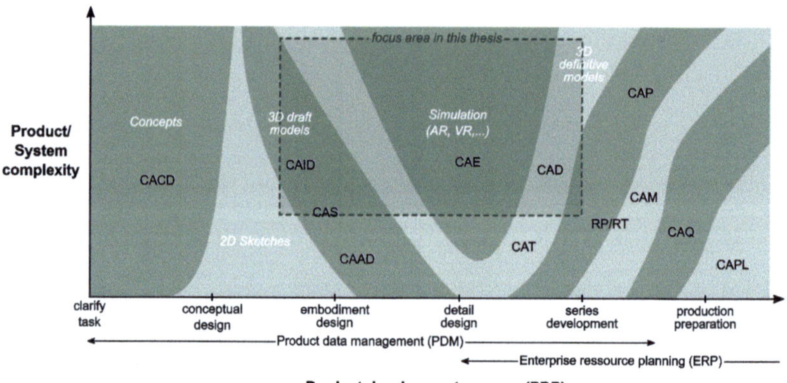

Fig. 2.9 Categorization of CAx systems according to their relevance in the PDP in relation to the respective complexity of a product based on [Vaj+18, p. 19; Wer+04, p. 1440] (compare Table 2.4 for abbreviations)

complex functions related to the physical product performance need to be described and optimized in the CAE environment to proceed with a detailed geometry design. Limitations in CAD tools become evident, especially when designing demanding parts, products, and assemblies, particularly for manufacturing processes that enable a high degree of design freedom, such as die-casting, AM, and others. Thus, in the context of this work, it is important to comprehend the composition of CAD tools to locate and identify limitations in engineering design tools.

Composition of CAD Tools

CAD systems consist of several functional elements to enable design engineers to visualize, modify, specify, structure, and, to a certain extent, optimize technically feasible geometries, parts, assemblies, and holistic products in a virtual software environment. In Fig. 2.10 six of those elements are exemplarily differentiated. First, the mathematically accurate and robust representation of the geometry is a fundamental task of any CAD system. Second, CAD systems consist of a variety of functions and features to enable the manipulation of the virtual geometry by the design engineer's input. Third, the virtual parts created in the CAD environment need to be structured and constrained according to their hierarchy in the product assembly. Fourth, to determine physical (e.g., total weight) or economic (e.g., estimation of manufacturing costs) product properties, material representations can be assigned to each solid of the product. Fifth, the capabilities of CAD tools are constantly evolving towards the increasingly accurate simulation of the product's physical performance with respect to the requirements it is subject to (CAE-functionalities). Sixth, in recent years functionalities towards algorithmic geometry generation or optimization, are being included in the scope of contemporary CAD (e.g., topology and shape optimization). This list is not exhaustive, as modern CAD tools also include additional functional elements not mentioned here. Nevertheless, those elements mentioned are essential to enable virtual engineering design freedom and will be further analyzed in upcoming chapters of this work.

Those functional elements can be structured in four layers which are explained in more detail below (Fig. 2.10). It is important to memorize that regardless of the scope and functionalities the key component of a CAD system still are design engineers themselves [KBF05]. As users of the CAD tool, design engineers are simultaneously enabling and limiting factors and thereby crucial for the completion of a purposeful design. In this sense, the layer descriptions of the system below are described from a design engineer's perspective. A description from a software developer's point of view might be technically more precise, particularly regarding the back-end. However, for the implications of this work, this level of detail is sufficient.

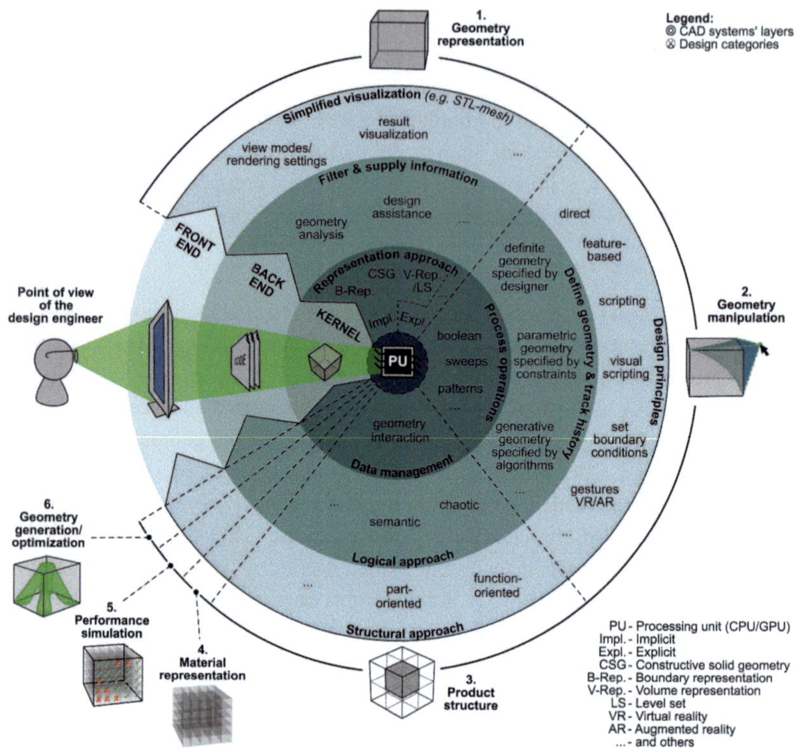

Fig. 2.10 Architecture and characteristics of CAD tools from a design engineer's perspective based on [Fuc+22]

1st Layer-Front-End: The first layer can be referred to as *front-end*. Front-end elements represent the only touch point between design engineer and CAD system, the so-called *human-machine-interface* (HMI) or *user-interface* (UI) that enables humans to interact with computers (Fig. 2.10 outer circular layer). One major function of the front-end is visualizing the 3D geometry, enabling an agile interaction often due to a simplified geometry file format. Besides the geometry visualization, the front-end also displays the geometry manipulation functionalities of the UI. These depend on the design principle incorporated in the system, enabling the design engineer to create and modify geometries in virtual 3D space. Additionally, an overview of the product structure is often displayed as a product tree. Therefore, the front-end can be understood as a mediator between the design engineer and the software incorporated in the *back-end.*

2nd Layer-Back-End: In the back-end are all those elements of the software embedded that are responsible for processing and providing necessary information to accomplish the desired task by the engineer. It puts the users' inputs into action, provides design assistance, and enables geometry analysis. The primary function is the mediation between the front-end and *geometric modeling kernel,* processing the information gathered by the frond-end and controlling the embedded geometry kernel accordingly. It tracks the design history and enables the design workflow while defining and generating the desired design output data for manufacturing (Fig. 2.10 second circular layer).

3rd Layer-Geometric Modeling Kernel: The main product attribute in the focus of CAD Systems is the virtual 3D geometry of a technical product. Mainly in charge of the representation of the geometry in the CAD system is the *Geometric Modeling Kernel* (GMK) (Fig. 2.10 third circular layer). Here, the product's geometry is mathematically determined in the most accurate, responsive, and stable manner. GMKs are also often capable of processing basic design operations, e.g., boolean operations. GMKs can be based on differing mathematical approaches for geometry representation. Depending on the design task, these approaches are a limiting factor in enabling the design of challenging and optimized geometries. Mathematical geometry representation approaches can be categorized into two groups: implicit and explicit representations.

> **Definition 9a:** *Implicit representations* use rules to check points in space for membership of a subset. It involves the definition of an indicator function that assigns the states true or false to arbitrary input points, depending on whether that point is inside or outside the subset. [Fuc+22, p. 6]

> **Definition 9b:** *Explicit representations* use rules to create points within a subset. A function is defined which creates points exclusively within the subset of space occupied by the part. [Fuc+22, p. 6]

In Table 2.5 common geometry representation methods are compared according to their performance within five different property types—geometric accuracy, computational stability, design freedom, memory efficiency, and property augmentability. Property augmentability refers to the possibility of calculating and storing further properties in addition to the pure geometric shape, such as local material properties or the mechanical deformation, by using a FEM simulation. This listing provides a general orientation and does not claim to be exhaustive. In summary, for solids, almost exclusively implicit methods are used in the field of CAD, and explicit methods with polyhedra are primarily used in the field of CAE. For a more comprehensive evaluation of digital geometry representation methods for different application purposes, please refer, e.g., to [BFR17].

Table 2.5 Overview of implicit and explicit representation methods for solids commonly implemented in geometric modeling kernels [Fuc+22]

	Representation Method	Geometric Accuracy	Computational Stability	Design Freedom	Memory Efficiency	Property Augmentability
Implicit	Implicit Functions	●	●	◔	●	○
	Spatial Decomposition	◔	●	◕	○	◕
	Polygon Meshes	◑	◔	◕	◑	◔
	Spline Surfaces	●	○	●	◕	◔
	Subdivision Surfaces	◕	◔	●	◕	◔
Explicit	Polyhedra Meshes	◑	◑	◕	◔	●
	Spline Volumes	●	○	●	○	●
	Subdivision Volumes	◕	◔	●	○	●

(○ = low; ● = high)

4th Layer-Processing Unit: Together, all three layers described above operate in accordance with each other, enabling the CAD system's complete functionality. All of them depend on and are limited by the computer's or server's processing unit and its capabilities (Fig. 2.10 core layer).

Conventional Computer-Aided Design Workflow

The focus of this section is set on the design engineers' workflow itself, as designer engineers are a crucial factor in the feasibility of highly optimized complex geometries. The design workflow of a design engineer is often highly specialized to the application context and particularly the respective manufacturing process in focus (e.g., sheet metal, casting, and others). Nevertheless, some patterns can be abstracted from engineering design workflows that, generally speaking, apply to a majority of design tasks. The upcoming description of an exemplary conventional design workflow is based on parametric-associative as well as history- and feature-based modeling tools only, as these are mostly used for technical applications (for further information, see [SD12]).

The product hierarchy in the automotive context is often set as follows. A *product* is a *system* composed of several individual *sub-systems*, again containing *components* that are compiled from different *assemblies* consisting of different *parts*. Nevertheless, for simplicity reasons in this work the two levels *component* and *assembly* are considered as one (Fig. 2.11 left).

The conventional design workflow illustrated in Fig. 2.11 is structured into five main steps and starts after the product planning phase is concluded (compare PDP in Fig. 2.2). This order means that the specifications, the functional scope as well as the fundamental concept decisions should already be completed before the detailed engineering design work in the CAD environment begins (compare Fig. 2.11).

In the first step of the design work product requirements, such as overall dimensional measurements and values to be met documented in the specification sheet, need to be translated into the design space. This step includes, e.g., the virtual location/orientation of the product, main package dimensions, assembly structure, part count, and specific non-design areas due to, e.g., technical principles, regulations, and ergonomics, altogether aspects of the *embodiment design*. The deliverables from this phase are *3D package models* and sometimes even freehand pre-sketches illustrating the first steps of the approach for the following *detail design*.

A product architecture often consists of structural components carrying subcomponents which in turn are built from assemblies comprising of different parts. Therefore, the detailed design phase in the following second and third step of the illustrated workflow in Fig. 2.11 often occurs on the part level of the product structure. Typical for conventional parametric associative modeling tools is to create necessary 2D reference geometries in the second step of the workflow. Those might be, e.g., planes and sketches (incl. points, lines, axles), often parametrically defined, describing basic profiles, topology, and shapes on which the 3D geometries are associatively built upon later on. Subsequent changes in those associated geometries and parameters significantly influence the downstream design history.

Based on the 2D digital reference geometries, the 3D part is modeled in the third step of the workflow. This step is nowadays primarily enabled by features both in surface and volumetric modeling environments. Additionally, inter-volumetric operations, also known as *boolean operations* can be applied to shape at least two independent closed volumes based on their intersections. The detailed part design is elaborated by iterating steps 2 and 3 of the workflow. Further, the interfaces to neighboring parts of the same assembly need to be considered. Results from this

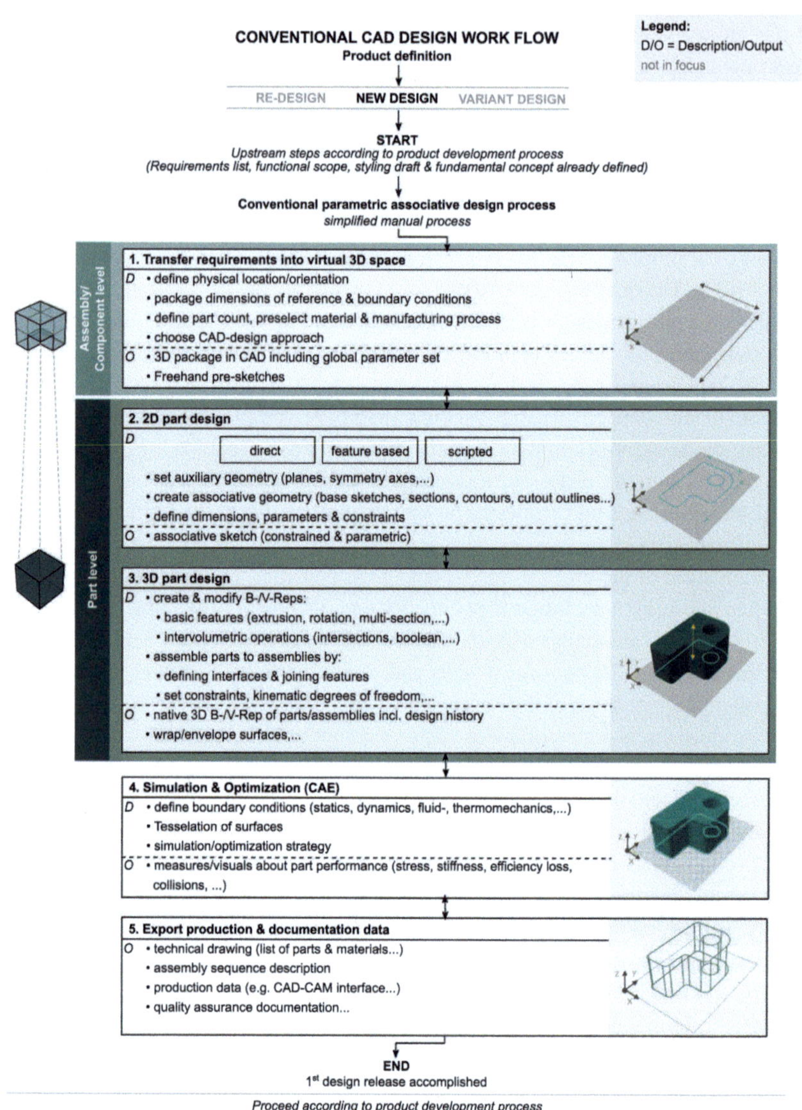

Fig. 2.11 Conventional parametric-associative CAD workflow described in a direct modeling fashion based on [Fuc+22]

phase are the native 3D design data based on volume (V-) or boundary representations (B-Rep) of the parts/assemblies drafts, including their design history and constraints.

The fourth step usually belongs to the context of CAE and addresses the virtual simulation and optimization of the suggested geometry in order to estimate its performance in service [KBF05]. In the last two decades, the relevance of this step increased significantly due to its gained validity and development costs saving potential, minimizing testing efforts, material usage, and part failures in use. Therefore, and due to its iterative character, CAE solutions are more and more merging with CAD software environments.

After the design drafts are validated and optimized through simulation and optimization, the design is finalized, the manufacturing data exported (CAM), and the design documentation concluded in the fifth step of the workflow (Fig. 2.11). The first design release is now accomplished, and the component proceeds in the development process with, e.g., prototyping and testing (Fig. 2.11 at the bottom). Additional design iterations are not unlikely as not all product performance aspects can be validated virtually beforehand yet. This workflow describes only the main stages of a generalized design task in CAD and does not claim to be holistically applicable to each specific design context.

To accomplish the modeling act by the design engineer described in the second and third step (see Fig. 2.11), three different established principles for geometry manipulation can be distinguished (see Fig. 2.12). Each has advantages and disadvantages and should be chosen wisely by the design engineer at the beginning of the detail design phase.

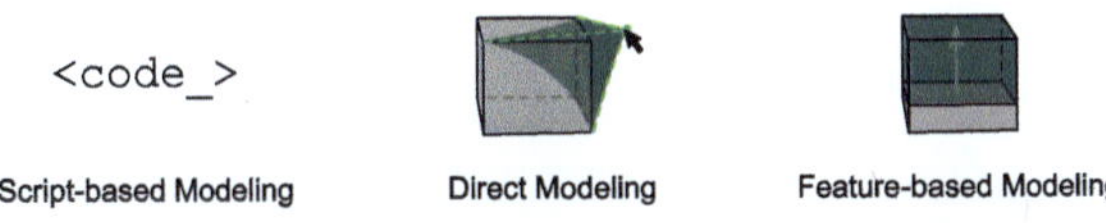

Fig. 2.12 Established conventional principles for geometry manipulation

Script-based modeling: Generating and manipulating virtual geometries in a coded manner. The coded script describes the subsequent geometrical operations and is compiled and represented by the CAD geometry kernel. Nowadays rarely used by design engineers. Particularly suitable for programming geometry generators that, once set up, create design variants based on varying input parameters automatically (Fig. 2.12).

Direct modeling: Generating and manipulating virtual geometries by mouse/touch-pad/touch-displays gestures (click, drag, pinch, etc.) directly visually affecting changes in the geometry in the desired fashion. Up today still strongly implemented design principle in mainstream CAD systems (Fig. 2.12).

Feature-based modeling: Generating and manipulating virtual geometries by means of preprogrammed features that enable to specify and implement a variety of design elements (e.g. extrusions, pockets, rotations, offsets, chamfers, etc.). This design principle has a substantial share in the functional scope of mainstream CAD systems and is often embedded supplementary to the direct modeling capabilities of the software (Fig. 2.12).

Those design principles demand specialized skills and often lead to intensive manual labor attempting to design highly optimized complex geometries. Depending on the design tool, the focus might differ between those design principles mentioned above. Table 2.6 lists and differentiates between exemplary typical engineering design and styling tools on the market.

Table 2.6 Exemplary overview of commercially available CAD and CAS tools based on [Fuc+22]

Category	
Software product	*Company*
Industry standard computer-aided-engineering design tools	
Creo, (Pro/Engineer)	PTC
NX, Solid Edge	Siemens
3DExperience, Catia, Solid Works, SFE Concept	Dassault Syst.
Fusion 360, Inventor, AutoCAD	Autodesk
Space Claim	Ansys
Inspire	Altair
...	...
Industry standard computer-aided-styling tools	
Blender	NaN, Blender Foundation
Rhinoceros 3D	Robert McNeel & Associates
Alias	Autodesk
...	...

With the fundamentals of structure and workflow of engineering tools, it becomes clear that CAD/CAE systems are complex means and indispensable in today's product development environment. Nevertheless, there are still limitations to those systems that come particularly into effect in designing high-complexity products.

2.1.3   Product Development for Mass Individualization

Design Methodology also faces challenges in developing increasingly relevant mass customizable products. This circumstance results from the satiation of basic needs in industrial societies increasingly shifting consumers' focus to personal fulfillment and individualism (compare *Hierarchy of Needs* according to MASLOW (1943) [Mas43]). With the resulting influence of the individualization trend on consumer behavior and needs, novel requirements for mass production are raised. Therefore, a turn from mass production over mass customization to mass personalization/individualization is observed from the early 80s on [Kor+15, p. 70] (Fig. 2.13). Thus, the diversification of demands is one of several external complexity factors for PD in a mass-production context [KG18, p. 21]. Further, increasing satiation of

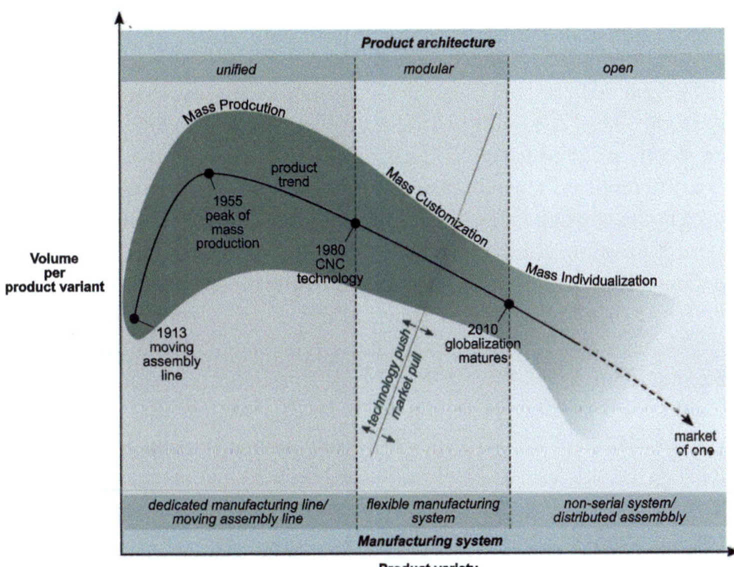

Fig. 2.13 Qualitative market development towards product individualization based on [Kor+15]

product demand caused by mass production in the automotive sector as well as in other industries over the last decades is being perceived [LRZ06]. The increasing diversification of model ranges enabled by group-wide platform and module strategies is one action to address small unaddressed market shares. Configuration complexity of off-the-shelf products with numerous specification options and additional accessories has been one strategy to address individual customer needs in automotive mass production. Unfortunately, this product strategy led to exponentially growing numbers of possible product variations, causing company-internal complexity gains and, therefore, costs. Thus, cost-saving variety management efforts are strictly minimizing configuration freedom for customers over the past years. An increased emotional value of products is, therefore, necessary to compensate for increasing overall similarity of products in quality and technical capabilities [Wag+16]. Novel product design and development methods must enable a superior compromise between product simplification and product individualization.

In this context several marketing terms like *personalization, customization* and *individualization* are often used synonymously and address the company-to-customer relationship with regard to communication, services and/or products [RT03, p. 36]. In the context of PD, it is crucial to differentiate those terms from each other, as they imply significant discrepancies in PDP complexity. Also, in the literature, definitions for those terms differ strongly from each other, making it challenging to identify and assort distinguishable term-meanings ([RT03, p. 36; Ves+05, p. 8; Zhe+17; KG18, p. 140; Hel+18]). Personalization and customization are mostly referred to and compared with each other in the literature. They are often differentiated regarding the active role in the company-to-customer relationship. On the one hand, personalization is mostly driven by companies that use customer information, either previously obtained or provided in real-time, to offer the most suitable product from its line-up, addressing the customer in a personalized fashion (Fig. 2.14 left). On the other hand, customization is often understood as primarily costumer-driven or customer-conducted product adaptions in order to meet personal requirements [Ves+05, p. 8]. This differentiation alone is not exhaustive for the automotive context. Here, the term vehicle customization incorporates an aspect of co-creation in which customers not only adapt their vehicle independently post-delivery, yet are at least involved in the product configuration. In contrast to vehicle personalization, in this sense of customization, customers and the company collaborate to configure the final product to a certain extent (Fig. 2.14 left). In other words, personalized products can be comprehended as *customer-specific* products and customized products as *customer-specified* products.

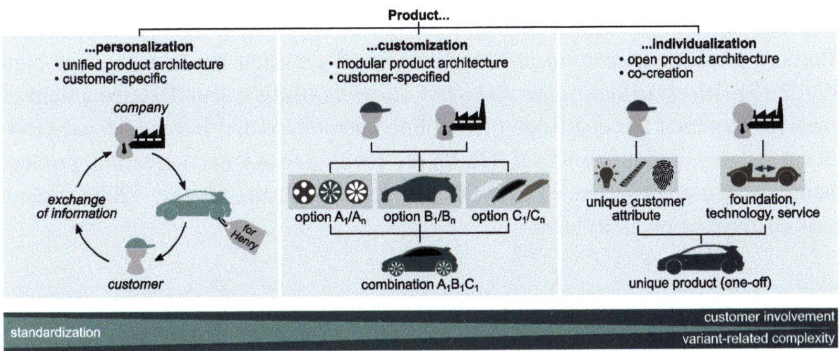

Fig. 2.14 Suggested term differentiation between *product personalization, customization and individualization* based on [RT03, p. 36; Ves+05, p. 8; LRZ06; Zhe+17; KG18, pp. 54, 140; Hel+18, p. 21; Kor21]

In the category of customer-specified products, it is also possible to distinguish further between customization and individualization. The degree of customer involvement in product development, specification, and production increases from one to another (Fig. 2.14 bottom). As a result, vehicles configured by customers themselves choosing and compiling different options of the company's predefined line-up to a certain product variant can be considered customized vehicles. The configured variant of the vehicle is aimed to meet customers' needs as well as possible. In the realm of individualization, though, unique attributes of the individual customer are taken additionally into account during product specification, in such a manner that unique products (one-offs) result rather than product variants (Fig. 2.14 right). [Kor21, p. 21; KG18, p. 255]

In conclusion, for the following considerations in this work, the level of customer involvement in product development and specification is chosen as the main differentiating factor for product personalization, customization, and individualization. This order also indicates the increasing company-internal complexity induced by the respective level of customer involvement.

The mass customization aspect is a key element considered in the product portfolio in the automotive industry. Fig. 2.15 emphasizes the increased model diversification of volume-based automotive OEMs in relation to the overall stable number of total annually sold units over the past decade. Additionally to the increasing model variety, it is common that, due to the high level of variant diversification enabled by the number of model configuration options, not every combinational possible vehicle variant is sold during the entire production period of the respective model

[KG18, p. 23]. As a result of the importance of customer-product identification, efforts to implement customization options in off-the-shelf product lines are high (Fig. 2.13). *Mass customization* and *mass individualization* also describe a field of research, targeting the challenge of enabling customized and individualized products at the mass production level [GL15, p. 7] with a focus on co-creation, product change, and user experience (UX) [Zhe+17]. LINDEMANN ET AL. (2006) define mass customization as follows.

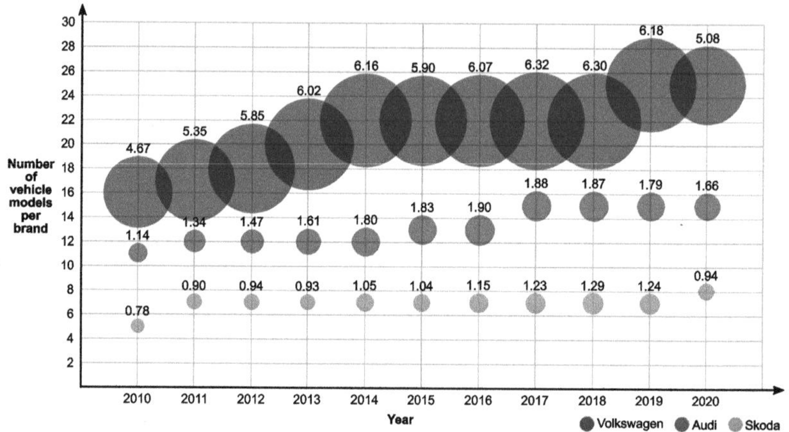

Fig. 2.15 Rise of vehicle model number per brand in the course of time in correlation with total sales volume in Mio./a (○) of exemplary OEM's based on annual reports (compare also [WFO09, p. 12])

Definition 10: *Mass Customization* (MC) – Strategic concept for deep integration of customers and their individual needs into the company and it's value chain. (translated from [LRZ06, p. 1])

Definition 11: *Mass Individualization* (MI) – is understood as an advanced level of MC with a focus on co-creation and based on individual customer attributes.

Table 2.7 contrasts the differences in the characteristics between special production (custom-made), multi-variant and individualized products. This comparison clarifies that to meet the MI trend (Fig. 2.13) it is necessary for a company to enable customer individual co-design in pre-planned structures based on a spectrum of products and processes. Here, the flexibility of products, processes, and corporal structures are key factors for a company's competitiveness.

Table 2.7 Differentiation between product customization and individualization based on and free translated from [LRZ06, p. 2]

Product type	Special Production	Multi-variant Products	Individualized Products
Example	Products from craftspeople/workshops	Automobiles, electronic devices	Consumer goods with individual components
Characteristics	• Development and production of individual products • Non-systematic and un-documented planning and execution of value-chain processes • Competitiveness through competences and experience (focusing strategy)	• Configuration and assembly of pre-developed and pre-assembled standardized modules • Distinct segmentation of product and production through modularization • Competitiveness through increased efficiency and productivity	• Individual design in pre-planned structures • Development of a spectrum of products and processes • Documentation and re-use of know-how • Competitiveness through flexibility of product, processes and company structures

In this context, the relation between variant-based individualized products and occurring internal complexity-related opportunity costs is particularly sensitive. As variant-related opportunity costs only occur lately in the PDP, opposed to the variant-related design decisions, it is challenging for companies to supervise and control variant-related internal complexity [LRZ06]. For this reason, modularization and variant reduction are ongoing trends in automotive companies in order to reduce combinatoric complexity and interrupt the vicious circle of product diversity [LRZ06; KG18]. This trend is contrary to MC, let alone MI in the automotive sector. Development approaches in this context aim to minimize this dilemma by suggesting efficient development approaches and modular product structures while finding new ways to consider individual customer attributes as input for product specification.

As mentioned above, the MC approach in the automotive industry is based on combinatoric of offered configuration options resulting in a high number of possible product variants. Nonetheless, the installation rates of those offered variants result in a pareto-distribution (5 % to 15 % of variants are installed frequently, the remaining 80 % of offered variants show installation rates of less than 1 %) [LRZ06, p. 8].

This relation implies that today's MC approaches generate corporate complexity in vain. Furthermore, according to PASEK ET AL. (2009) customer satisfaction does not necessarily only improve with increasing product variety.

To compensate for this paradigm, LINDEMANN ET AL. (2006) suggest, among other authors, an adapted product development structure based on MI instead of MC and on product specification instead of configuration (Fig. 2.16). Here, a transition from variant-specific product development to the development of a product spectrum is proposed. In this approach, it is not intended to reach a specific milestone in the PDP, in which all variants of the product range are fully defined. As in MI, the product specification is based on unique customer attributes, yet it is not feasible and intended to foresee each possible product variant. In MI, product diversity shall not be avoided but enabled by specifically flexible product-, process, and corporate structures. Therefore, the development phase is rather focused on the adaptability of the product structure and the technically reasonable as well as feasible specification intervals of the product spectrum.

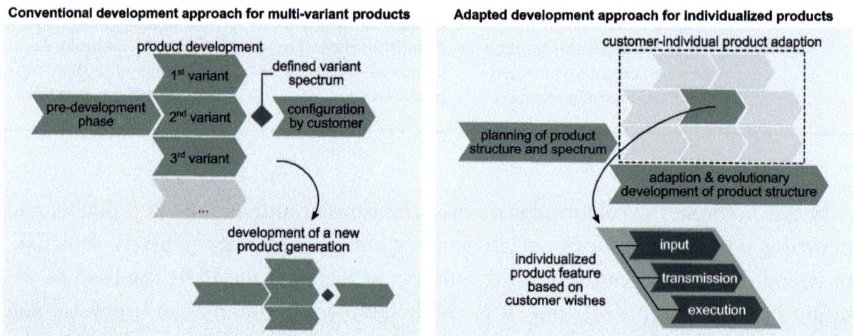

Fig. 2.16 Comparison of simplified conventional and individual development approaches based on [LRZ06, p. 13]

According to the literature, there are at least two ways of enabling co-creation effectively and user-friendly [Zhe+17]. The first one is by computer-aided configuration systems, often based on web applications, widely used in the context of multi-variant consumer products [Hel+18]. The other is a postponement method by designing products with built-in flexibility or adaptability, enabling customers to specify the product according to their needs during usage [Zhe+17]. Further, computer-aided configuration tools can be distinguished into two slightly differing categories, *configuration tools* and *specification tools*.

> **Definition 12:** *Configuration Tool* – Configuration tools also known as *product configurators* [Li+18] or *sales configurators* [GL15, p. 6] are knowledge-based systems that incorporate the combinatoric logic of the product structure, illustrate the respective configuration options to the customer in a manageable and descriptive fashion as well as gather the customer requirements as input. [LRZ06; Li+18; GL15]

Particularly in the automotive industry, the logic in the back end of such configurators is highly complex. Configuration options are often offered as bundles to counteract the configurational complexity for customers and the company. Even though state-of-the-art configuration tools are getting increasingly sophisticated, incomprehensible combinatoric limitations often lead to a negative customer experience [GL15].

> **Definition 13:** *Specification Tool* – Compared to configuration tools, specification tools also incorporate the product architecture and bridge the gap between customer requirements and the final product. Contrary to product configurators, specification tools do not only rely on the selection of configuration options by customers but rather focus on the translation of unique customer attributes into unique product designs within the individualization scope [LRZ06, 28 ff.].

Following the definitions 12 and 13, configuration tools are mostly applied for MC, and specification tools are mostly used for MI [LRZ06]. Even though the difference between both tool types seems marginal, the development approach and effort for each respective product structure might differ significantly. Fig. 2.17 contrasts product configurators with specification tools in exemplary manner.

Fig. 2.17 Exemplary non-automotive and automotive configuration and specification tools based on [LRZ06]

2.2   State of Vehicle Conception and its Upcoming Challenges

According to MACNEILL AND CHANARON (2005) the three main drivers of trends in the automotive industry are: global competition, legislation, and consumer demand [MC05, p. 88]. Nowadays, STOLFA ET AL. (2020) affirm that new technologies and business models, climate goals, environmental and health challenges, and societal changes and changes in consumer behavior sum up to over 80 % of the drivers for change in the automotive industry in the contemporary literature [Sto+20, p. 13]. Further KALMBACH ET AL. (2022) summarize contemporary challenges in automotive industry into five points:

- Real customer focus
- Autonomous driving
- Connectivity and digitization of vehicles
- Electrification of power trains
- Shared mobility

Considering the saturation of the automotive market in developed countries those challenges result in at least two paradigm changes for the industry. The first is the transition from the internal combustion engine to battery electric vehicles, and the second is the transition of OEMs from vehicle manufactures to providers of autonomous mobility. Vehicle conception (VC) depends on current development trends to introduce the right product at the right time. The increasing number of derivatives offered in the market is one indicator of OEMs' effort to exploit additional market niches (Fig. 2.15) [Fra09b]. Due to those interdependencies named above, it is important to emphasize the environment levels starting from *abstract/global* to *specific/vehicle* that influence vehicle conception shown in Fig. 2.18.

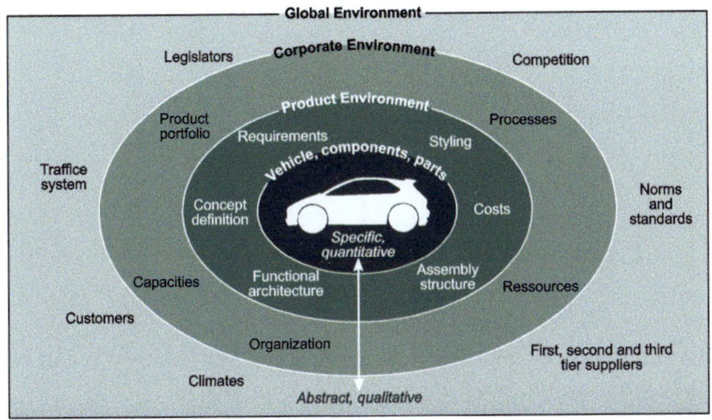

Fig. 2.18 Environmental levels to be considered during vehicle conception based on [Fra09b; Neh; Lu+18]

This section introduces four aspects of VC. First, Sect. 2.2.1 gives a brief introduction to vehicle physics to comprehend the basic functional principles. Second, vehicle compositions are explained to convey the technical sections of a car in Sect. 2.2.2. Third, in Sect. 2.2.3, the VC process is introduced to clarify causes and effects. Fourth, subsection 2.2.4 emphasizes fundamental conflicts of objectives in VC, as these are particularly relevant for the findings of this work.

2.2.1   Vehicle Physics

The physical circumstances in which a vehicle operates imply fundamental correlations for its engineering design. e.g., the relation between different driving resistances indicates the importance of respective properties for specific vehicle types. Eq. 2.1 shows the sum of driving resistances F_{DR} of the vehicle.

$$F_{DR} = F_{Acc} + F_{RR} + F_{AD} + F_{RG} \tag{2.1}$$

The driving resistances can be distinguished in to four main forces, *acceleration resistance* F_{Acc}, *rolling resistance* F_{RR}, *aerodynamic drag* F_{AD} and *slope forces* F_{RG} with RG representing the rolling gradient. With exception of the aerodynamic drag F_{AD} all resistance forces F_{Acc}, F_{RR}, and F_{RG} are strongly dependent on the vehicles overall mass m (Eq. (2.2), (2.3), and (2.5)). F_{AD} is mainly influenced by the relative velocity v_r^2, the frontal cross section A, and the aerodynamic drag coefficient c_d of the vehicle (Eq. (2.4)) [BS13, 50 ff. KS14].

$$F_{Acc} = m \cdot a + \sum_n^i \frac{J_i \cdot \ddot{\phi}_i}{r_i} \tag{2.2}$$

$$F_{RR} = m \cdot g \cdot f_r \cdot \cos \alpha \tag{2.3}$$

$$F_{AD} = c_d \cdot A \cdot \frac{\rho}{2} \cdot v^2 \tag{2.4}$$

$$F_{RG} = m \cdot g \cdot \sin \alpha \tag{2.5}$$

The overall vehicle power necessary to reach a certain velocity v can be obtained with the following correlation (Eq. (2.6)) [BS13, p. 50].

$$P = F_{DR} \cdot v \tag{2.6}$$

Crucial factors influencing the driving resistances and therefore the vehicle's efficiency are the vehicle's mass m, rolling resistance f_r and aerodynamic drag coefficient c_d, as well as the drive train efficiency η_u. Eq. (2.7) considers the driving resistances and their crucial factors mentioned above to determine the vehicle's fuel consumption for a certain distance B_e in dependence of changes in velocity-over-time for vehicles with internal combustion engines [BS13, p. 52]. To enable the comparability of vehicle models' energy efficiency and emissions, standardized velocity-over-time driving cycles (e.g., "Worldwide Harmonized Light Vehicles Test Procedure" – WLTP) can be considered [Tut+13].

$$B_e = \frac{\int b_e \cdot \frac{1}{\eta_u} \left[\left(m \cdot f_r \cdot g \cdot \cos \alpha + \frac{\rho}{2} \cdot c_d \cdot A \cdot v^2 \right) + m(a + g \cdot \sin \alpha) + B_r \right] \cdot v \cdot dt}{\int v \cdot dt} \tag{2.7}$$

Figure 2.19 illustrates the relation between driving resistances and vehicle speed. Those dependencies clarify the main influences regarding energy efficiency properties of the vehicle and therefore need to be considered in an early stage of the vehicle conception. To determine the holistic energy demand of a car, all losses due to energy conversion and drive train friction, as well as all energy-consuming auxiliary users, such as air conditioning and others, need to be considered. Besides the driving resistances, additional principles of vehicle-physics impact vehicle design. e.g., correlations with origin in driving dynamics require specific chassis properties and imply rigidity constraints for the vehicle body. These constraints are, e.g., also related to vibration- and acoustic comfort, track performance, and longevity.

Each sub-aspect mentioned in this section is a complex topic in itself and therefore cannot be clarified in detail within the scope of this work. Nevertheless, it is important to emphasize that physical principles have a profound impact on vehicle conception setting the limits of what is feasible. For further information on this topic consult [Pri11; BS13; Fri13; BF18; Hak18; PS21].

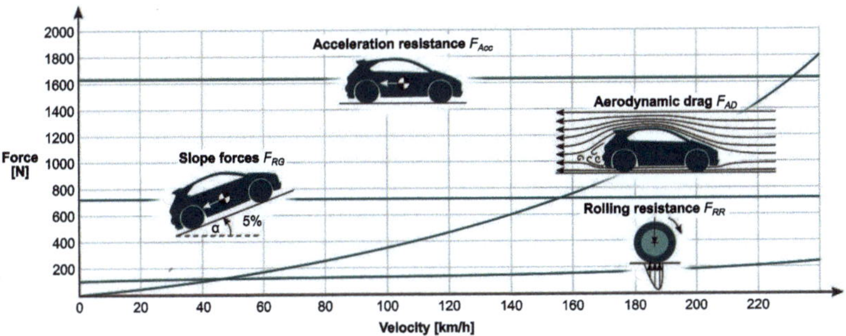

Fig. 2.19 Exemplary driving resistances in relation to vehicle velocity considering aerodynamic drag, acceleration, gradient and rolling resistance for a medium-sized passenger car based on [Hak13, p. 137; Sch13, p. 144]

2.2.2 The Composition of a Vehicle

Conventionally, at least four technical sections can be distinguished to describe the architecture of a vehicle as illustrated in Fig. 2.20: *vehicle body*, *chassis*, *drive train* and *electrics/electronics*. Each vehicle part is technically and organizationally

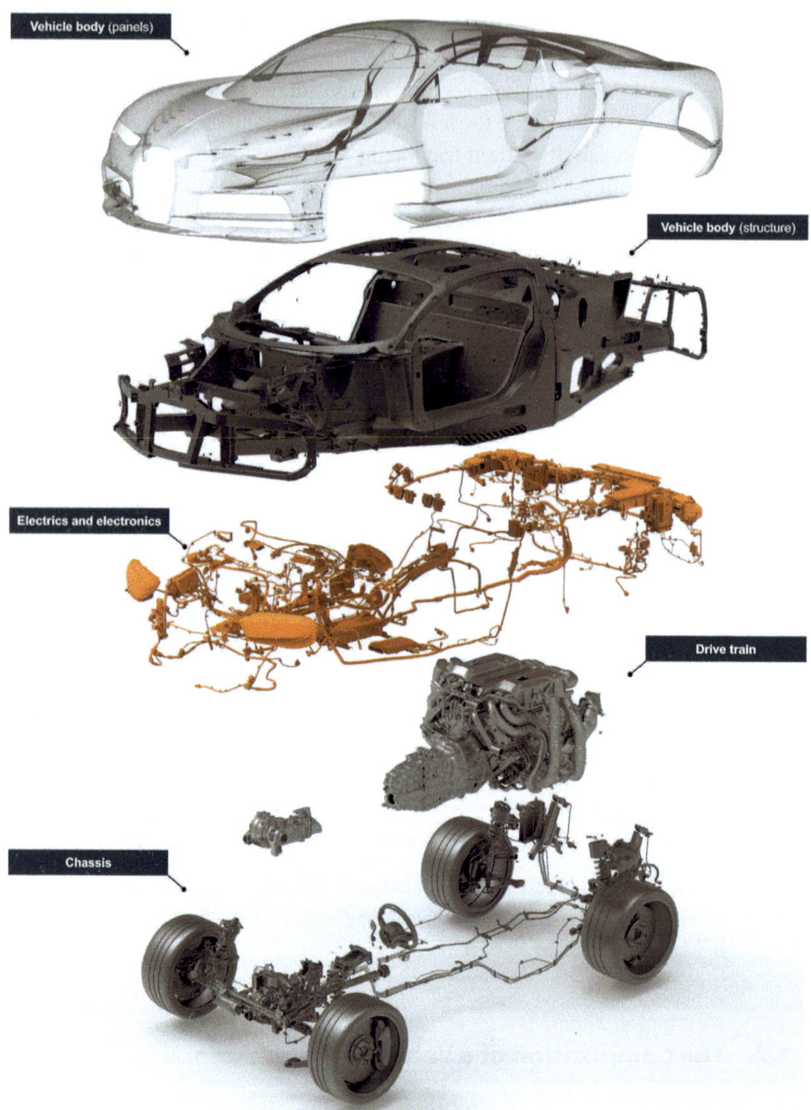

Fig. 2.20 Exemplary overview of the main technical sections of a vehicle as first level of its functional architecture (by courtesy of Bugatti Engineering GmbH)

subordinated to one of these technical sections. Therefore, the disciplinary organizational structure of automotive development departments often correlates to these respective sections. Each technical section, is focused on its functional scope and omits the variance of possible design solutions.

- **Body:** The vehicle body contains the load-carrying structure of the vehicle. This technical section is, among others, responsible for the structural integrity of the car considering driving and crash-related loads. Additionally, the vehicle's exterior components, such as panels, doors, lids, hoods, glass, interior components, such as seats, dash panel, and lining, are subordinated to this section. Therefore, the vehicle's exterior and interior styling is also strongly related to this technical section.
- **Chassis:** The suspension, including, among others, springs and dampers of a vehicle, is crucial for its comfort and driving stability and is considered the primary function of the vehicle's chassis. Further, the tires, steering, braking, and stability control systems of the vehicle are also main components of this technical section.
- **Drive train:** The drive train is responsible for powering and propelling the vehicle. All parts and assemblies necessary to do so, e.g., energy storage, engine, and transmission, belong to this technical section.
- **Electrics/electronics:** Lights, sensors and actuators, control units, and all the wiring necessary for electronic and mechatronic components are part of this technical section including the power supply system vehicle.

Nevertheless, considering the ongoing developments in the automotive industry, additional previously subordinated areas are gaining importance. For example, the software-based functional scope and the perceptive/sensory vehicle functions are taking over significant development capacities for upcoming vehicles and must be recognized as central elements of its functional architecture.

Most common vehicle types are composed of the four technical sections mentioned above. One major point of differentiation between different vehicle types is their topology. A simplistic way to approximate the overall topology of a vehicle category is the arrangement of one or more boxes as representatives of the most dominant volumes. Accordingly, vehicle categories can be distinguished as e.g. one-box-, two-box- or three-box-design (Table 2.8) [BS13, p. 137].

As implied in Table 2.8, besides their respective topology, vehicles can be further categorized into different classes addressing varying segments of the market. Each vehicle class results from the combination of fundamental vehicle characteristics that relate to a specific group of target customers. Fig. 2.21 shows different ways

Table 2.8 Differentiation of vehicle categories approximating their topology characteristics with boxes based on [BS13, p. 137; Kuc12, p. 10]

	One-Box-Design	**Two-Box-Design**	**Three-Box-Design**
Description	The entire vehicle is perceived as one volume (one box). Nevertheless there are separations between passenger cabin, storage space and technical components, such as drive train, engine and transmission	Division of the vehicle in to two section/volumes. One volume, mostly in the front of the vehicle, consists of the engine bay. The second volume consists of the passenger cabin and the luggage compartment	Conventional segmentation of the vehicle. Separation of technical, passenger and luggage compartment. An increased variability between passenger area and luggage compartment is observed in three-box configurations
Variants	• Vans • Small car concepts	• Station wagons • Sport utility vehicles (SUVs) • Compact class vehicles	• Limousines • Classic coupés • Convertibles • Roadster
Advantages	• Low footprint • Efficient utilization of space • High seating position (good overview in traffic)	• Variability between passenger area and luggage compartment possible	• High acoustic comfort due to compartment separation • In favor of good passive safety properties (luggage isolated) • Beneficial for high vehicle body stiffness properties
Dis-advantages	• Increased vehicle height and front cross-section • High center of gravity unfavorable for driving dynamics	• Big interior space disadvantageous for acoustic comfort • Coefficient of drag strongly dependent on rear angle	• Low space variability • Stalling edge needs to be clearly defined

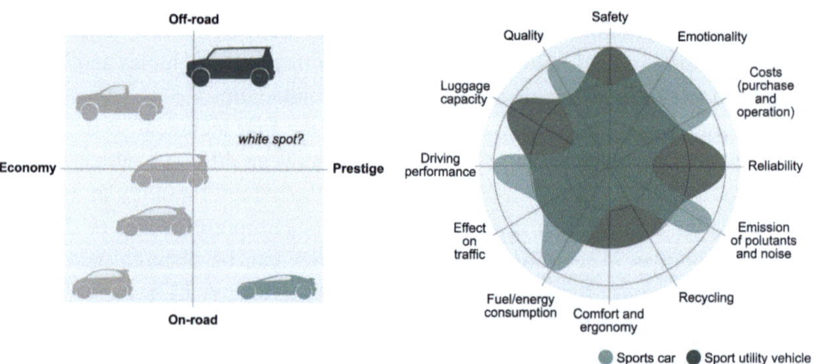

Fig. 2.21 Combination of vehicle characteristics distinguish vehicle classes to address diverse target customer groups based on [MWG09, p. 53; GK10, p. 101; PS21, p. 1279]

of visualizing differing vehicle properties between vehicle segments. It becomes clear, that vehicle classes can have contrary properties in comparison with each other, e.g. *sport utility vehicles* (SUVs) and *sports cars* regarding *fuel consumption*, *driving performance*, and *comfort* (Fig. 2.21 *left*), yet still satisfying their respective target group of customers. For further information on vehicle segments and their characteristics, consult [PS21, p. 138; MWG09, 13 ff.].

The technical sections of a vehicle introduced in Fig. 2.20 can be realized following differing approaches. Exemplary approaches for the realization of the *vehicle body* are explained in more detail, as those are in focus of the elaborations of this work. At least two main differing design approaches exist to achieve the structural integrity of a vehicle, the *body-on-frame* and the *body frame integral* architecture (Table 2.9). The *ladder frame* design and the *triangulated tube structure* design belong to the body-on-frame type of structural architecture and are known for their simplicity, robustness, and the separation of structure and shell. On the downside, their high package demand and low functional integration potential limit their field of application mostly to commercial vehicles (ladder frame) or race cars (triangulated tube structure) only. The *space frame*, *unibody* and *monocoque* structural design methods are assigned to the body and frame integral architectures. In this particular order, the fundamental structural elements of the design change increasingly from frame-based to shell-based structures. The unibody is the most common in today's consumer vehicle production, as it is the most cost-effective design method in mass production. The monocoque structural design approach incorporates the highest lightweight, bending-, and torsional-stiffness potential and is therefore common in high-performance vehicles.

For further specifics on the remaining three technical vehicle sections chassis, drive train, electrics/electronics, please consult the respective literature (e.g., [HEG07; MWG09; Hen13; BF18; PS21]

Table 2.9 Overview of the vehicle structural architectures based on [SS08; SK15]

	Conventional vehicle body design methods	Advantages	Disadvantages
Body-on-frame architecture	**Ladder frame**	• Simplicity • Robustness • Separation of structure and external shell • High bending stiffness	• Torsional sensitivity • High package demand • Low functional integration potential
	Trellis frame	• Robustness • Simplicity due to separation of structure and external shell • High bending stiffness • High torsional stiffness	• Sensitivity selective bending loads • High package demand • Low functional integration potential • High manufacturing effort and costs
Body frame integral architecture	**Profiles and shells** (Space Frame–Audi AG)	• Particularly robust • Partially separation of structure and external shell • Very high bending stiffness • High torsional stiffness • Low package demand • High functional integration potential • High lightweight potential	• Elevated manufacturing effort • High requirements on joining processes and corrosion prevention
	Shell design (Unibody–VW AG)	• Efficient and cost effective in mass production • High bending stiffness • Very high torsional stiffness • Relatively low material costs • Low package demand • High functional integration potential	• High tooling costs • Limited lightweight potential due to manufacturing process and costs
	Tub design (Monocoque–Porsche AG)	• Load specific design • Optimized weight distribution • High functional integration potential • Very high lightweight potential • Very high bending stiffness • very high torsional stiffness	• Manufacturing complexity • Design of fiber reinforced parts • Joints and connection points • Manual labor intensive

2.2.3   Vehicle Conception and Design Process

One objective of the conception phase of a vehicle is to determine its most suitable composition to fulfill all necessary requirements (Fig. 2.18) as well as possible. The origin of each vehicle concept is an initial product idea based on the anticipated market demand. After the product planning phase, the product idea is elaborated from coarse to detailed during vehicle conception. The outcome of this phase is the technical description of the respective vehicle concept with all relevant documentation.

> **Definition 14:** *Vehicle Concept* – A vehicle concept is an engineering design draft of a product idea that outlines its technical feasibility. The draft consists of a composition of fundamental parameters, modules, and components that influence basic vehicle properties and characteristics [BS13, p. 130]. The design and definition of the following elements are thereby in focus:
> - Vehicle type, shape, structure, and necessary variants,
> - Number of passengers, space required by ergonomics,
> - Storage space and necessary volumes (e.g. tank or battery),
> - Main dimensions,
> - A concept for engine and drive train.

Figure 2.22 displays an overview of the vehicle conception process. It becomes clear that the process can be subdivided into the *concept definition* and *concept elaboration*. During the definition phase, the alignment of the vehicle concept and determination of relevant requirements is conducted. In the elaboration phase, the technical and the styling concept are derived by refining rough drafts into detailed designs. The technical and the styling concept are developed parallel to each other, negotiating possibilities and limitations.

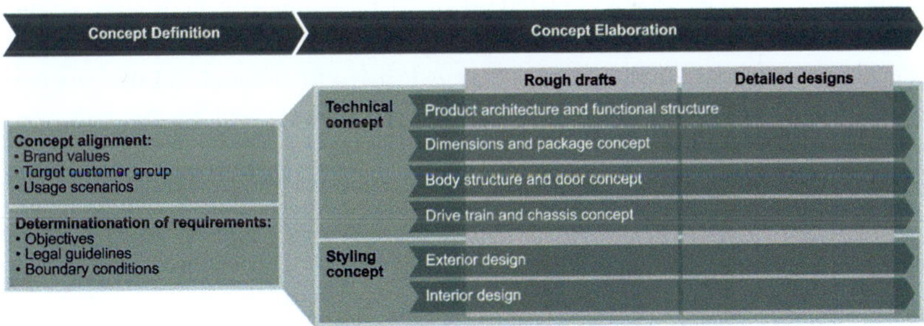

Fig. 2.22 Overview of the vehicle conception process based on [Pri11]

Nonetheless, the vehicle conception process is subordinated to the PDP introduced in Fig. 2.2 and Fig. 2.3 and starts in the early phases *product planning*, *product definition*, and ends with the *concept freeze* where *design detailing* begins. Non product-based organizational units of a company such as research, pre-development department, and innovation management carry the responsibility to generate, develop and implement innovative ideas and technologies at the right place and time in the company. They are a valuable source of input for product- or process-related innovations to consider during vehicle conception. As implied in Fig. 2.22 the entire vehicle conception process is extensive and requires interdisciplinary teamwork across organizational units. [PS16]

Phases of the Vehicle Conception and Design Process

In the following two distinct phases of the vehicle conception process are briefly explained to clarify key elements that are crucial for the final embodiment of a vehicle. The VC process begins with the *concept definition* phase, that can be subdivided into *concept alignment* and *determination of vehicle requirements* steps.

Vehicle concept alignment: The guiding values of a brand are the foundation of every vehicle concept in the making. Those values affirm ethical principles and legal guidelines and define the brand's orientation. These are values with which potential customers identify themselves either consciously or unconsciously. They are valid for all corporate activities and thus impact the vehicle conception. Those values are crucial for long-term mutual consent between the company, society, and government. [PS21]

The values of the society and their development tendencies also need to be considered to align a vehicle concept successfully. Those are interpreted based on trend analysis and foreseeable disruptive influences. Here, futurologists may help assess society- and technology-related trends. Determining target customer groups in society is necessary to align a vehicle concept precisely. It can be helpful to consult categorization models such as *Sinus Milieus* [Sin16] to comprehend the structure of the respective society. This way, customer desires, needs, and life circumstances are taken into account during concept creation [Foe09].

Further, a vehicle-related scenario is developed based on brand values, trends in society and technology, as well as a defined target customer group. A scenario describes a fictional use of the prospective vehicle in various situations in the daily life of the targeted customer group. A story is derived with suitable personas, in which customer needs are emphasized, and how the prospective vehicle meets those needs is clarified. Additionally, the story also approximates the intended operating conditions of the vehicle, enabling the determination of product requirements as the next consecutive step [MWG09]

Determination of vehicle requirements: After defining major boundary conditions during vehicle concept alignment, it is crucial to determine and document the vehicle requirements in a list of requirements or product specification sheet. These are fundamental for the following concept elaboration and can be consequently checked for fulfillment. Basic requirements can be derived from the boundary conditions of the vehicle concept, e.g., range, number of passengers, necessary storage space, and the vehicle's maximum speed. Additional requirements result from the global product environment (Fig. 2.18). Especially considering vehicle properties regarding safety, emission of noise and pollutants, vehicle conception underlays the legislation and consumer protection organizations. Further, requirements are classified according to their priority. Mandatory requirements, such as those prescribed by law, are classified as fixed requirements. Additional categories can be, e.g., maximum-, minimum-, interval- or nice-to-have-requirements. Each of the levels illustrated in Fig. 2.18 imposes requirements upon the subordinated level. If those are insoluble, fixed requirements must be adjusted iteratively, or if conflict of objectives arise, compromises need to be found. [FG13] Fig. 2.23 emphasizes the complexity of requirement categories and points out interdependencies resulting in conflicting objectives.

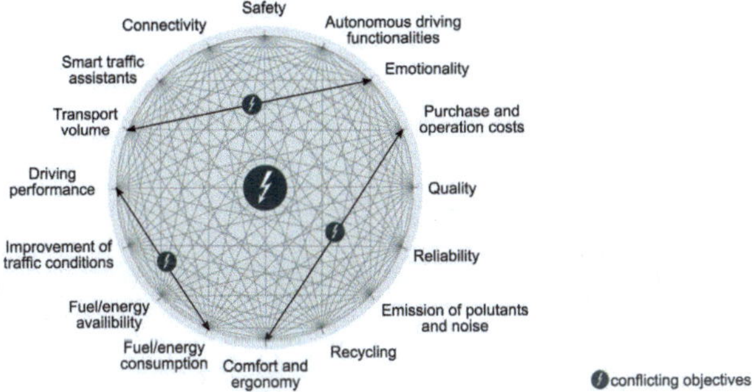

Fig. 2.23 Network of automotive requirement categories based on [PS16, p. 12] and exemplary conflicting objectives based on [WFO09, p. 125]

The vehicle *concept elaboration* phase is the second main phase of the VC process in which rough and later detailed drafts for the technical and the styling concept of the vehicle design are developed.

Vehicle concept elaboration: In the vehicle concept elaboration begins when the concept definition phase is concluded. The technical and the styling concept are developed simultaneously yet closely coordinated in this phase (Fig. 2.22). This way, it is ensured that the technical and the styling concept match when unified [PS16]. As repetitively stressed in VDI 2221, it is crucial to ensure the fulfillment of the list of requirements or the specification sheet throughout this elaboration [VDI93]. The vehicle's product architecture and functional structure are necessary to oversee all necessary sub-solutions of the concept. The main interface between engineering and styling is the *technical package* of the vehicle. The package defines all relevant single and continued dimensions and constraints (Fig. 2.24) and all components to be considered in the design space. The dimension concept is broadly influenced by the anthropometry and ergonomics of potential occupants. Additionally, design- and non-design spaces can be derived from the package in such a way that interfaces between engineering and styling as well as between exterior and interior are clarified.

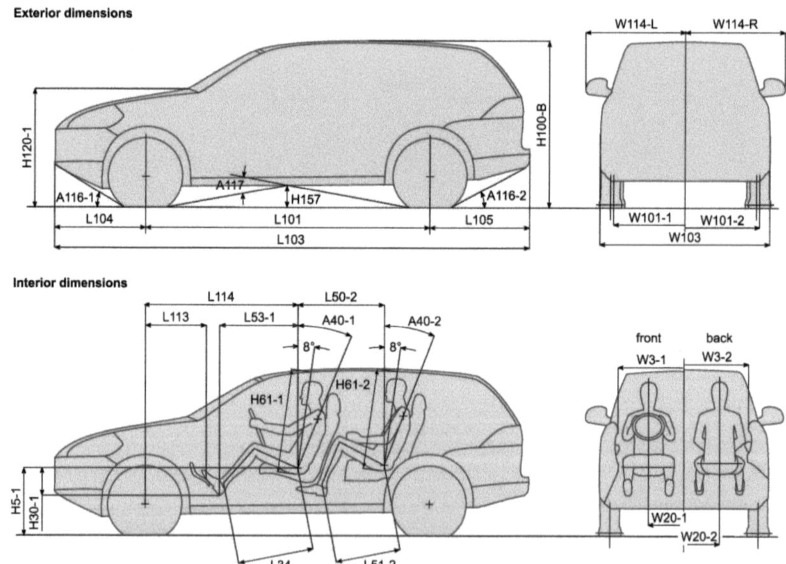

Fig. 2.24 Exemplary exterior and interior continuous dimension concept content of a technical package of a passenger vehicle based on [BS13, 133 ff.]

Definition 15: *Vehicle Package* – The package is a technical draft of the vehicle gradually elaborated during the development phase. This technical draft aims to ensure the technical feasibility of the planned product and verify the dimensional interplay of all assemblies and components, thereby considering all requirements derived from...

- customer relevant and legal,
- environment and safety related,
- stylistic,
- economic,
- technically functional,
- production and maintenance relevant, and
- quality assuring ...

... aspects, combined to obtain the best possible holistic vehicle design solution. In other words, differing conflict of objectives, design space claims, and functional dependencies are considered in such a manner that a geometrically and physically compatible composition of all vehicle components, the so-called *package*, results [Bra97; BS13, p. 131].

Engineering design of sub-concepts, e.g., vehicle body, doors and closures, drive train, and chassis follows next [Pri11; Fra09b]. The entire vehicle conception process is supported by CAD and CAE software tools. Styling and engineering models are built digitally, enabling further alignment, detailing, and optimization of the design draft further down the line. Those phases described above are well incorporated into the PDP of an OEM (Fig. 2.3). Through simultaneous engineering, manufacturers can develop vehicles from the concept idea to their start of production in less than four years. The corporate module strategy is crucial for an international OEM such as Volkswagen AG. In order to keep variant complexity manageable, a particular focus is set on module management throughout the development process. Furthermore, the majority of vehicle components are provided by tier-one and tier-two suppliers, which is why suppliers are also aligned with the PDP [WFO09, pp. 40, 144]. Due to their extended product portfolio, international OEMs face the challenge of enhancing development efficiency and simultaneously enabling a diversified product line-up. Therefore, vehicle conception is elaborated routinely according to strict specifications. As many requirements of prior models as possible are carried over to the next product generation (Fig. 2.3) and only necessary vehicle modifications are conducted keeping proven vehicle properties and minimizing development effort [Web09].

Due to numerous involved technical disciplines, departments, and employees involved in developing vehicle concepts and process step iterations (necessary because of new findings), the conception process is time-consuming and cost intensive. Furthermore, a long project lead time might result in a decreased degree of product innovation. Alternative manufacturing technologies that accelerate conception and development phases (e.g., AM) are therefore in particular focus during vehicle conception. The following subsection discusses conflicting engineering objectives and their relevance for vehicle conception in more detail.

2.2.4   Conflicts of Objectives in Vehicle Conception

Sect. 2.2.3 and Figs. 2.18 and 2.23 imply the complexity of multi-level requirements to be considered during vehicle conception. Additionally, it becomes clear that those requirements might be mutually dependent, causing conflicts and thereby reducing engineering design freedom during conception [Pri+13, 167 ff., 179]. In the engineering design context, terms such as, e.g., conflicting constraints, requirements, demands, goals, conflict of objectives, and trade-offs are often used to express conflicts between differing engineering objectives. EILETZ (1999) defines the overall term *conflicts of objectives* as follows (*translated from* [Eil99, p. 15]):

Definition 16: *Conflict of Objectives* – A conflict of objectives means the mutual, substantiated dependency of at least two objectives of each other, whereby the (gradual) fulfillment of one objective compulsorily impedes or inhibits the fulfillment of the other. A conflict of objectives consists thereby of the following three components:

- At least two objectives are affected
- The objectives are related to each other regarding one common entitlement object
- One specific manifestation of the entitlement object is responsible for the conflict of objectives

One example for conflicting of objectives in automotive design is the strong relation between requirements on passenger comfort while entering and exiting the vehicle, and the fulfillment of safety requirements regarding side crash performance. The initial challenge while managing conflicting objectives is to identify and overview those conflicts within complex systems. To do so FRANKE (2009) elaborates on the need for relation systems that describe dependencies in between objectives. Further, PRINZ ET AL. (2013) emphasize that, to overview conflicting objectives in a complex system, such as a vehicle, computer assistance is crucial. Fig. 2.25 suggests an approach on how to methodically cope with conflicting objectives.

Fig. 2.25 Approach to cope with conflicting objectives in engineering design based on [Eil99]

Besides the simple prioritization of conflicting requirements, the following four listed solution strategies for coping with conflicting objectives are mentioned and explained in detail in the literature [Pri+13, p. 187]:

- Sensitivity-analysis of individual vehicle parameters/characteristics;
- Optimization of the already established vehicle as a system;
- Derivation of differing vehicle variants to diversify prioritization of the set of requirements according to specific demand;
- Use of innovative technologies in order to solve conflicting objectives (new degrees of freedom).

The last item mentioned above is essential for this work, as AM is presumed to be an innovative technology with the potential to mitigate engineering compromises derived from conflicting objectives during vehicle conception. With this in mind, AM is further introduced in the following section (Sect. 2.3, Sect. 2.3.1).

2.3   Design for Additive Manufacturing - Potentials and Restrictions

In this section, the fundamentals of Additive Manufacturing (AM) are summarized (Sect. 2.3.1). The focus is on Design for Additive Manufacturing (DfAM). This particular field of research addresses methods and approaches that enable design engineers to consider AM restrictions and implement product- and process-related AM potentials during product/vehicle development (Sect. 2.3.2). Subsequently, the contemporary understanding of those respective AM potentials and their resulting value propositions are explained in further detail (Sect. 2.3.3).

2.3.1 A Brief Introduction to Additive Manufacturing Technologies

Continuous improvements in the technological readiness of toolless and material-addition-based manufacturing processes enable novel opportunities in end-product manufacturing. Engineering design restrictions derived from conventional manufacturing processes (e.g., processes based on material subtraction, casting, forming, forging, and other tooling-dependent processes) represent a technical feasibility limit for design engineers and need to be learned by hard. However, those restrictions only partially apply to the manufacturing technologies in the focus of this section, unleashing unconventional approaches to technical design. 3D-Printing (3DP), Generative- (GM), Rapid- (RM) Direct-Digital- (DDM), or Additive Manufacturing (AM) are common terms used to summarize these technologies. 3DP is commonly used in an informal context. Elsewhere AM is established as an umbrella term in the literature (comp. [VDI14; Gib+21)]. Even though 3DP is the most popular term, AM is primarily and synonymously used in this work. AM is defined as follows in the literature [VDI14, p. 3; Kum18, p. 7; GKT19, 2 ff. Gib+21, pp. 1–3]:

> **Definition 17:** *Additive Manufacturing* (AM) – AM technologies are manufacturing processes in which physical parts are built up through automated material addition of volumetric elements such as voxels or cross sections directly derived from a digital 3D-geometry file. A typical characteristic of AM processes is their independence of product-specific tools and preparations (toolless manufacturing).

Origins

AM finds its origins in the early 1980s, in the development of the *Stereolithography Apparatus* [GRS16, p. 37]. In the same decade, other processes such as Laminated Object Manufacturing (LOM), Selective Laser Sintering (SLS) and Fused Deposition Modeling (FDM), joined Stereolithography (SLA) forming a group of tool-less manufacturing processes. Based on MEINERS (1999) and GIBSON ET AL. (2016)

Fig. 2.26 summarizes some of the most important milestones in the development history of AM technologies.

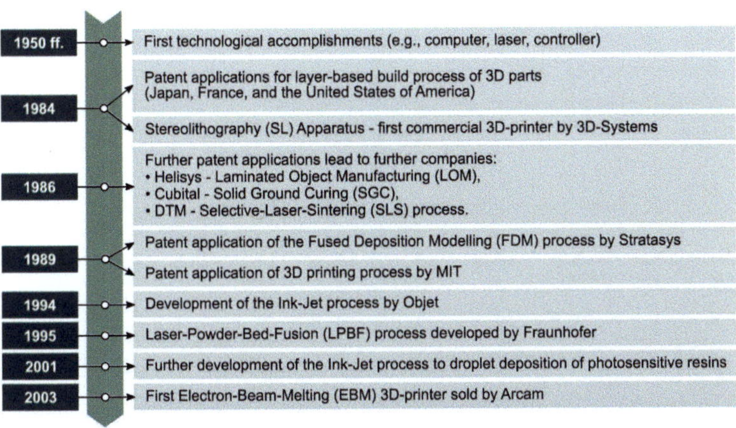

Fig. 2.26 Milestones in the history of AM based on [Mei99; GRS16]

Application Areas

Mainly used to manufacture solid models with sufficient accuracy to facilitate design-, styling-, package- or assembly-related engineering decisions, those processes were of particular use during early stage product development substantiating the term *Rapid Prototyping* (RP). With increasing dimensional accuracy and improving material properties, the fields of application broaden according to the three F's (form, fit and function), enabling soon the tool-less manufacturing of functional prototypes that partially fulfill product requirements [GRS16, p. 3] (see Fig. 2.27).

Due to their limited large-scale production capabilities, most AM processes are considered for applications with small lot sizes. Besides prototype parts, this is mainly the case for additively manufactured production equipment and tools, establishing the category *Rapid Tooling* (RT) [GRS16, p. 437; GKT19, p. 14].

With improving technology readiness, AM is increasingly used to produce end products establishing the overarching term *Additive Manufacturing* even further. Furthermore, a fully automated process chain from the digital model to the final product is a critical objective in developing the technology. Therefore, the term *Direct Digital Manufacturing* (DDM) is also closely related to AM, not least because of the flexibility attributed to AM processes. Based on GEBHARDT ET AL. (2019)

Fig. 2.27 illustrates the term relations concerning their areas of application in the AM context. Here, it becomes clear that prototyping and manufacturing are separated from each other [VDI14, p. 5]. Further, in the automotive industry, prototypes are classified into subcategories, from A-samples to C-samples. Those prototype categories are subject to increasing requirements, each challenging AM technologies even further. Fig. 2.27 also illustrates that AM technologies are not limited to prototyping-only applications anymore, yet are being applied to the entire product life cycle.

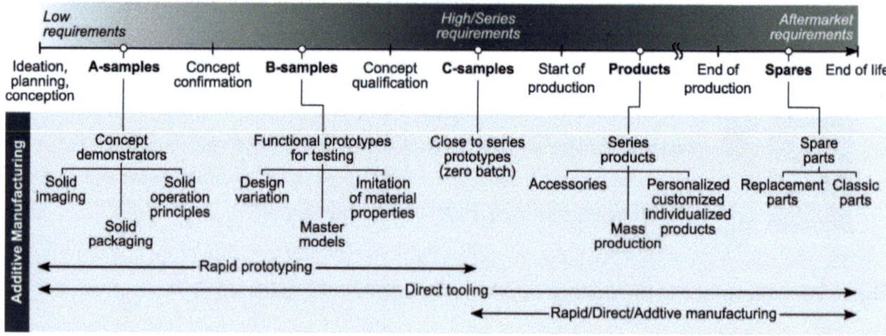

Fig. 2.27 Overview of application areas of AM based on [Geb16, p. 14; GKT19, p. 14; BG20, p. 974]

Classification of Manufacturing Processes

Due to the vast application of AM technologies in the PLC, it is important to classify AM as a technology on equal terms with other manufacturing technologies. There are different approaches to categorize manufacturing technologies. According to the German industry standard DIN 8580 (2003), manufacturing processes are classified according to six main categories (Fig. 2.28). Those categories differ from each other regarding three main objectives. Those objectives can be either the generation or the modification of an original shape or the alteration of material properties. The processes to modify a shape are further differentiated into maintaining, reducing, or enhancing cohesion in the part. [DIN03]

Due to the diversity of AM technologies, it has been challenging to classify AM processes into the approach mentioned above in the past. So some processes were suggested to be classified as forming, a mixture of bonding and cutting, or coating technique [Kum18, p. 9]. Therefore, BENDER AND GÖHLICH (2020) classified AM

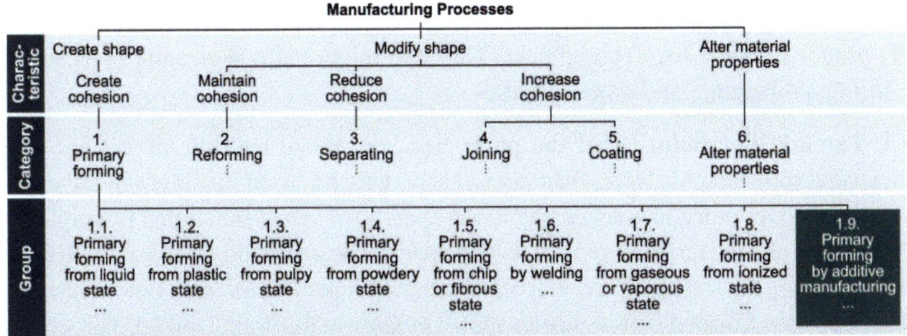

Fig. 2.28 Classification of AM in overall manufacturing process overview based on [DIN03, p. 7; DIN20, pp. 9, 11]

into the *special processes* category [BG20, p. 973]. Nevertheless, with increasing technological readiness level AM has now been classified as the ninth official sub-group in the *primary forming* category in the updated draft version of DIN 8580 [DIN20].

General Additive Manufacturing Process

Even though AM processes differ strongly regarding physical operation principles and the form and aggregate state of the raw material, most of them share a similar process chain. As shown in Fig. 2.29, the typical AM process can be divided into two phases, the virtual and the physical phase, or three phases, pre-process, 3D-printing, and post-process. The virtual phase belongs to the pre-process and starts already with planning (I.), conception (II.), embodiment, and detail design (III.), providing a digital production-ready 3D-CAD-model of the part to be additively manufactured. These three phases belong to DfAM and are analyzed in more detail in Sect. 2.3.2. The additional steps IV to VIII are described in the following (based on [VDI14; GRS16; Kum18; GKT19]):

IV. Data preparation: From here, the CAD file is mostly exported as tessellated STL-file and has to be strategically oriented into the virtual build chamber of the 3D printer. If necessary, temporary structures are added to support, e.g., overhanging part sections. The geometry is then sliced into cross-sections or divided into voxels, generating the mathematical information necessary for the specific 3D printing process. This step is also known as „slicing".

V. Machine preparation: Leaving the solely digital level of the AM process, the 3D printer needs to be prepared next. This step might include cleaning, feedstock refilling, preheating, and other activities.

VI. The additive build-up of the part: Now, the actual build-up process of one or several parts begins. Here, the material is added and fused in a layer-by-layer or voxel-by-voxel procedure according to the specific physical functional principle of the process and the targeted geometry of the parts. The part build-up is accomplished by generating the last layer/voxel. In most cases, a stepped outer part contour results. Therefore, the finished part is only an approximation of the digital model. This effect is also known as the *staircase effect*. Thinner layers or smaller voxels can achieve higher surface quality/accuracy by increasing the overall build time.

VII. Part extraction and clean-up: In this step, the part is extracted from the build chamber. This step might include depowdering, washing, building platform detachment, and other substeps. In most AM processes, removing the abovementioned support structures that might still be attached to the part demands particular effort.

VIII. Part finishing: Depending on the requirements, additional finishing steps might be necessary to finalize the product and meet customer demands regarding the overall quality impression, including surface roughness, color, and dimensional accuracy. To conclude the production, steps such as cleaning, sanding, shot peening, painting, and many other refining activities might be necessary.

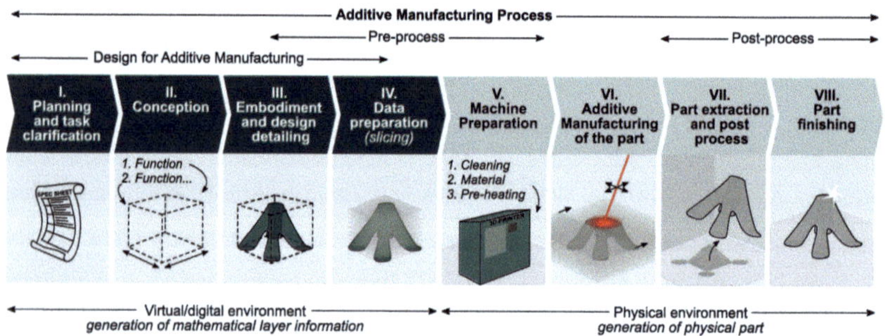

Fig. 2.29 General overview of the additive manufacturing process chain based on [GRS16, p. 45; GKT19, p. 4; Kum18, p. 10; Fuc+22, p. 4]

Benefits and Challenges of the Technology

There are several motives for the application of AM technologies. Those motives can be differentiated into two primary objectives, to achieve *product-related* or *process-related benefits*.

> **Definition 18:** *Product-related AM benefits* – Product-related benefits include all contributions enabled by AM that optimize the performance, the functional scope or the value-for-money-relation of a product. The customer can directly perceive these benefits.

> **Definition 19:** *Process-related AM benefits* – Process-related benefits include all effects caused by AM that influence the design, development, manufacturing, provision, sales, and storage process of a part, component, assembly or entire product positively. These benefits are directly perceived by the OEM and only indirectly perceived by the customer (e.g., through a shorter time-to-product).

Table 2.10 summarizes typical product- and process-related benefits and challenges related to AM mentioned in the literature. It becomes clear that most product-related AM benefits originate from additional design freedom resulting from the independence of constraints imposed by conventional manufacturing technologies. Even though AM processes are subject to their restrictions, additional design freedom enables improvements in, e.g., material and design space usage, efficiency losses, aesthetics, a.o. This way, parts, components, assemblies, and entire products can be optimized.

Besides the product-related AM benefits, challenges in this area must be faced. Due to the layer-by-layer manufacturing principle of many AM technologies, heterogeneous instead of homogeneous material behavior, particularly mechanical properties, result. This occurrence often limits the structural reliability of AM parts. Also, the resulting staircase effect often leads to poor surface quality, among others. Considering process-specific thermal influences and shrinkage effects, the dimensional accuracy of AM parts is still challenging today (Table 2.10).

Table 2.10 Product- and process-related AM benefits and challenges based on [Gao+15, p. 68; GRS16, pp. 404, 411; Tho+16, 5 ff. Kum18, pp. 17–22] (*this list is not exhaustive*)

Product-related AM ...	Process-related AM ...
... benefits	**... benefits**
• Dismiss constraints imposed by conventional manufacturing processes • Freedom in design and manufacturing of complex geometries enables, e.g.: 　• Part consolidation 　• Functional integration 　• Multi-material parts 　• Embedded objects and electronics 　• Interlaced assemblies with moving parts 　• etc... (see Subsec. 2.3.2) • Lightweight design • Reduction of design, installation, packaging space • Heat transfer improvement • Less efficiency losses • Improvement in aesthetics • Increased reliability • Individualization of geometries	• Cost reduction effects regarding: 　• Production costs 　• Assembly costs 　• Storage costs 　• Transport costs 　• Operational costs • Development time reduction • Sustainability improvements
... challenges	**... challenges**
• Material properties 　• Material heterogeneity 　• Structural reliability • Surface quality (staircase effect, down skin) • Dimensional accuracy (shrinkage or thermal internal stresses) • Individualization vs. mass manufacturing	• Process stability and reproducibility • Building scalability vs. layer resolution • Material variety • Build chamber size • Economic efficiency and profitability 　• Printing speed and batch-characteristic limits profitable lot sizes 　• Low process integration and degree of automation lead to high personnel costs 　• High raw material costs 　• High influence of nesting rates on part costs • Limited design knowledge as well as the suitability of conventional CAD and CAE tools for complex geometries • Material and process certification and quality control standards • Intellectual property

Process-related benefits can also be obtained concerning the process characteristics of AM technologies. This kind of benefit often leads to cost-reduction effects in the company. For example, production costs can be decreased due to the independence of AM processes of product-specific tools. Through part consolidation also, assembly costs can be minimized. *Production-on-demand* enabled by AM can minimize storage costs for spare parts or their respective tools. AM processes' digital and flexible nature also favors *production-on-location* diminishing transportation and shipment costs. Additionally, the targeted deployment of functional AM prototypes as early-stage replacements for conventionally manufactured prototypes can reduce development time. Through, e.g., material savings in structurally optimized parts and the resulting energy savings in its use also, sustainability benefits can be achieved.

Nevertheless, AM-process are confronting many challenges in manufacturing end-use applications. For example, process stability and reproducibility need to be improved. Material variety, certification, and quality control standards need to improve further. To be addressed, many processes' economic efficiency and profitability are strictly limited due to their printing speed and batch-characteristic. Additionally, limited design knowledge and the suitability of conventional CAD and CAE tools for complex designs are hindrances to fully leverage this manufacturing technology's potential.

Overview of AM Processes

As described in the definition of AM (Definition 17), processes based on a material addition principle for object generation are considered part of the AM process category. Thus, many different manufacturing processes are being subordinated to AM. A common approach to organize those AM processes is to structure them according to the physical state or form of the starting material, followed by their actual functional material fusion principle. Fig. 2.30 gives an overview of the most common, established, and upcoming processes clarifying their interrelations [VDI14, p. 8; Hua+15, p. 2; Kum18, p. 13; Gib+21, pp. 34, 386].

According to the diversity of AM processes, their process-related benefits and restrictions differ strongly. Therefore, some processes are more suited for certain applications than others. For example, some processes can print multi-material parts (PJM, FLM, LA, CS), and others might not need limiting support structures (LS, 3DP, MBJ). Further, some processes are rather suited for applications demanding high mechanical, thermal, or chemical requirements than others (LS, LBM, EBM, DED, CS). Nevertheless, other processes achieve high dimensional accuracy and advanced surface quality instead (SL, PJM, MJM). Besides discrepancies in the characteristics of the producible parts, these processes also differ strongly in the acquisition and operating costs of the respective manufacturing system. In the literature, these AM processes showed in Fig. 2.30 are compared extensively with each other assessing their specific advantages and disadvantages (compare [PG98, pp. 1281–1282; Gao+15, p. 69; Tof+18, p. 25; Kum18, p. 15; GKT19, 33 ff. WTH20, p. 3]).

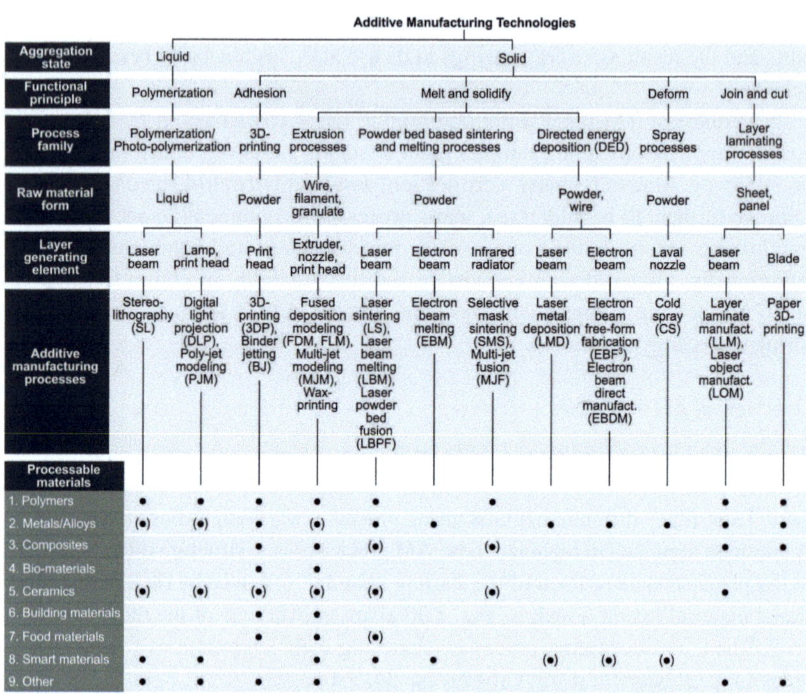

Attribute	SL	DLP / PJM	3DP / BJ	FDM / MJM	LS / LBM / LBPF	EBM	SMS / MJF	LMD	EBF / EBDM	CS	LLM / LOM	Paper 3D-printing
Aggregation state	Liquid	Liquid	Solid	Solid	Solid	Solid	Solid	Solid	Solid	Solid	Solid	Solid
Functional principle	Polymerization	Polymerization	Adhesion	Adhesion	Melt and solidify	Melt and solidify	Melt and solidify	Melt and solidify	Melt and solidify	Deform	Join and cut	Join and cut
Process family	Polymerization/ Photo-polymerization	Polymerization/ Photo-polymerization	3D-printing	Extrusion processes	Powder bed based sintering and melting processes	Powder bed based sintering and melting processes	Powder bed based sintering and melting processes	Directed energy deposition (DED)	Directed energy deposition (DED)	Spray processes	Layer laminating processes	Layer laminating processes
Raw material form	Liquid	Liquid	Powder	Wire, filament, granulate	Powder	Powder	Powder	Powder, wire	Powder, wire	Powder	Sheet, panel	Sheet, panel
Layer generating element	Laser beam	Lamp, print head	Print head	Extruder, nozzle, print head	Laser beam	Electron beam	Infrared radiator	Laser beam	Electron beam	Laval nozzle	Laser beam	Blade
Additive manufacturing processes	Stereo-lithography (SL)	Digital light projection (DLP), Poly-jet modeling (PJM)	3D-printing (3DP), Binder jetting (BJ)	Fused deposition modeling (FDM, FLM), Multi-jet modeling (MJM), Wax-printing	Laser sintering (LS), Laser beam melting (LBM), Laser powder bed fusion (LBPF)	Electron beam melting (EBM)	Selective mask sintering (SMS), Multi-jet fusion (MJF)	Laser metal deposition (LMD)	Electron beam free-form fabrication (EBF[3]), Electron beam direct manufact. (EBDM)	Cold spray (CS)	Layer laminate manufact. (LLM), Laser object manufact. (LOM)	Paper 3D-printing
Processable materials												
1. Polymers	•	•	•	•	•		•				•	•
2. Metals/Alloys	(•)	(•)	•	(•)	•	•		•	•	•	•	•
3. Composites			•	•	(•)		(•)				•	•
4. Bio-materials			•	•								
5. Ceramics	(•)	(•)	(•)	(•)	(•)		(•)				•	
6. Building materials			•	•								
7. Food materials			•	•	(•)							
8. Smart materials	•	•		•	•	•		(•)	(•)	(•)		
9. Other	•	•	•	•							•	•

Fig. 2.30 Overview of AM processes organized according to their primary functional principle including processable materials based on [VDI14, p. 8; Hua+15, p. 2; Tof+18, p. 26; PNR18, pp. 3877–3879; Kum18, p. 13; Gib+21, 34, 384 ff., 395 ff. BS21, p. 7]

Similarly to the diverse functional principles for material fusion, the types of materials possible to process in AM are also diverse. Fig. 2.30 shows the material categories being additively manufactured in commercially available systems. Among others, polymers, composites, metals and alloys, ceramics, sand, and paper belong to the most common categories [VDI14, p. 8; PNR18, p. 3879; Gib+21, 384, 395 ff.]. Relatively to metallic materials, a wide range of polymer materials is already available. WIESE ET AL. (2020), e.g., analyzed up to 130 differing polymers available and processible with established AM processes. Unfortunately, based on their compatibility with automotive requirements, only a few are already suitable for automotive applications [WTH20, p. 11].

Various metallic materials are additively manufacturable, too. VAFADAR ET AL. (2021) give a comprehensive overview of contemporary types of materials available for each metal-based process [Vaf+21, pp. 4–7]. Further, NGO ET AL. (2018) describe respective metallic materials in more detail [Ngo+18, 175 ff.]. In conclusion, despite the variety of materials available for AM processes, many have yet to meet requirements for applications in high-standard industries, such as automotive, aerospace, or healthcare. In particular, material longevity properties concerning chemical, light, heat resistance, and fatigue do not yet always meet quality standards.

Exemplary Applications in the Automotive Context

Even though numerous AM applications exist in different industries such as aerospace and health care industry [Che+17; Gao+15; Tho+16], Table 2.11 gathers examples from the automotive context. It becomes clear that polymer and metal parts with high geometrical and functional complexity are already feasible. Nevertheless, it is crucial to consider that those components differ strongly in technological readiness level and might not yet be suitable for series applications with a high level of requirements or high lot sizes (compare with *application areas of AM* in Fig. 2.27). Table 2.11 shows, on the one hand, explicit vehicle components such as body- (e.g., space frame node), chassis- (e.g., wheel, suspension upright, and brake caliper), drive-train- (e.g., drive train and wheel carrier, and transmission housing), and interior-related (e.g., dashboard air vent frame, seat bracket, seat back-rest, and gear lever knob, center console dial). Therefore, the state-of-the-art shows that AM is being considered in all technical sections of a vehicle. On the other hand, AM might also lead to benefits being applied in the production context as equipment (e.g., badge fixture). Also indicated in Table 2.11 is the diversity of product- or process-related benefits which each of those applications is aiming for (e.g., weight reduction, production cost reduction, realizing aesthetics designs not yet conventionally feasible (Table 2.10)). In conclusion, the application of the technology is driven by many different factors, and these are further introduced in the following Sect. 2.3.2 and 2.3.3.

Table 2.11 Exemplary overview of additively manufactured automotive applications

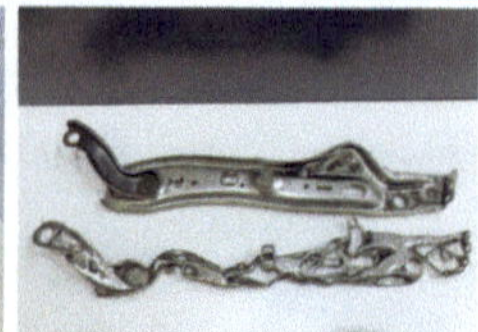

Description	**Automotive wheel**	**Race car suspension upright**	**Stamped/3D-Printed Seat Bracket**
Source	by HRE Wheels, GE Additive	by URE, Additive Industries	by SLM Solutions
Proc. – Mat.	Laser Powder Bed Fusion – Titanium	Laser Powder Bed Fusion – Titanium	Laser Powder Bed Fusion – Steel
Benefit	Improved aesthetics	Weight reduction	Weight reduction
Status	Prototype	Functional prototype	Prototype

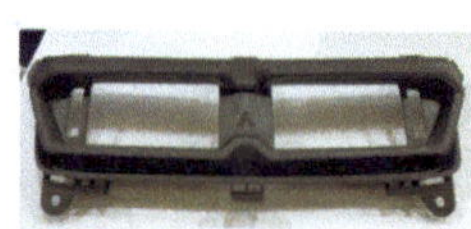

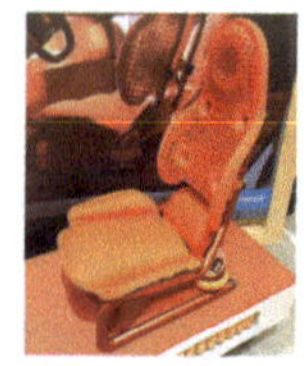

Description	**Dashboard air vent frame**	**Passenger seat back-/headrest**	**Badge Fixture**
Source	by Lamborghini, Carbon	by Citroen, BASF	by Volkswagen,
Proc. – Mat.	Digital light processing – Epoxy resin	Selective laser sintering – Polymer	Fused Deposition Modeling – Polymer
Benefit	Cost reduction of small lot sizes	Part consolidation, improved aesthetics	Cost reduction of small lot sizes
Status	Small series part	Concept demonstrator	Equipment

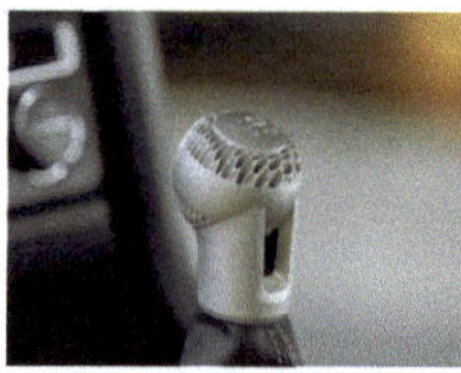

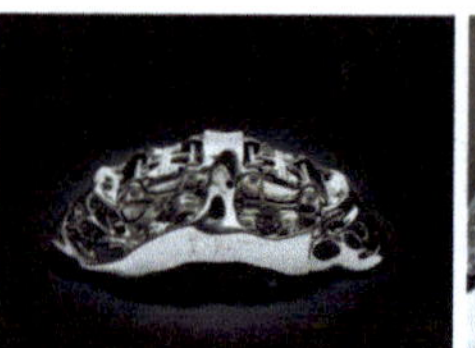

Description	**Gear Lever Knob**	**Break Caliper**	**Drive Train and Wheel Carrier**
Source	by Volkswagen, HP [HP18]	by Bugatti Eng. GmbH [Bug18]	by Renishaw GmbH
Proc. – Mat.	Metal Binder Jetting – Stainless steel	Selective Laser Melting – Titanium	Selective Laser Melting – Aluminium
Benefit	Cost reduction of small lot sizes	Weight reduction	Part consolidation, weight reduction
Status	Prototype	Functional prototype	Functional prototype

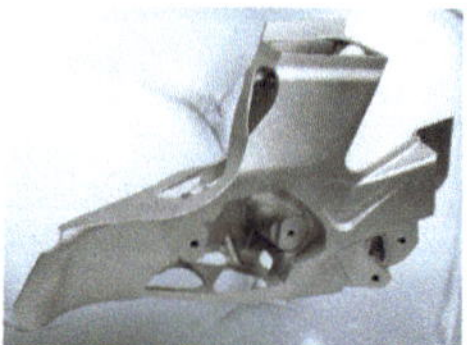

Description	**Space Frame Node**	**Transmission Housing**	**Rotary Center Console Dial**
Source	by Audi AG [Eib16]	by Porsche AG, Robert Hofman GmbH	by Bentley Motors [Ben22]
Proc. – Mat.	Selective Laser Melting – Aluminium	Selective Laser Melting – Aluminium	Selective Laser Melting – Gold
Benefit	Weight reduction	Weight reduction, part consolidation	Improved aesthetics
Status	Functional prototype	Functional prototype	Small series part

2.3.2   Design for Additive Manufacturing

Along with AM's advantages in efficiently producing small lot sizes locally and on-demand [GRS10; Geb16], its design potentials and restrictions are being explored in the field of research *Design for Additive Manufacturing* (DfAM) [Tho+16; BCP12; Kum+18; Kum18; LLK20; Fuc+20]. Besides AM potentials, e.g., *reduced time to product/market, assembly cost reduction*, the realizable design complexity is inspiring engineers to develop alternative part geometries resulting in innovative technical solutions often portrayed in showcase applications (Table 2.11). Nonetheless, AM technologies present design engineers with particular challenges in applying design potentials since many potential combinations exist, often increasing development complexity.

The DfAM research field addresses this problem by providing knowledge in the form of different general [Tho+16] and application focused [Ros07a; Tho+16; KFV23] methods and tools. Further, it is also essential that a change in mindset occurs in the early conception and design phases of product development. Otherwise, the design freedom of AM processes cannot be fully leveraged [BCP12; GRS16]. Traditional design patterns based on manufacturing constraints and potentials with respect to conventional processes must also be adapted [Ros07b]. In the research field DfAM extensive design and manufacturing process knowledge is analyzed, elaborated, structured, and provided through the literature. Here databases and methodologies are suggested to enable design engineers to apply AM potentials while considering technology-specific restrictions [Fuc+22].

There are several different definitions of DfAM in the literature. Early contributions e.g. from ROSEN (2007) or GIBSON ET AL. (2010) derive DfAM from the general context of *Design for Manufacturing and Assembly* (DfMA) and argue the need for an AM-specific body of knowledge additionally to DfMA [Ros07b, p. 403; GRS10, 401 ff.]. From this perspective DfAM is the synthesis of shapes, sizes, geometric mesostructures, and material compositions and microstructures to best utilize manufacturing process capabilities to achieve desired performance and other life-cycle objectives [Ros07b, p. 403; Ros07a, p. 586].

Meanwhile, DfAM evolved further, and more exhaustive definitions have been introduced. For example, the definition introduced by KUMKE (2018) also emphasizes the necessity to consider not just AM-related design freedom but also process-specific restrictions (e.g., necessary support structures, heterogeneous material properties, staircase effect on surfaces, post-processing efforts). Therefore, the following definition is used in this work.

> **Definition 20:** *Design for Additive Manufacturing (DfAM)* – DfAM describes methods and means that address the methodological design process holistically and assist with the identification and application of AM-specific design potentials as well with the consideration and implementation of AM-specific restrictions (free translated from [Kum18, p. 42]).

Further, also a framework distinguishing DfAM into two categories *DfAM in the strict sense* and *DfAM in the broad sense* has been suggested by KUMKE ET AL. (2016). The former consists of approaches concerning the design process (e.g., implementing design potentials and considering design rules). Latter focuses on the up and downstream specifics (e.g., part and process selection, manufacturability analysis) [KWV16].

DfAM Process

The general DfAM process is well understood and numerously reported in related work [Gib+21; Geb16; KWV16; Bel+18]. Fig. 2.29 illustrates the AM process steps in the primary focus of DfAM approaches and methods. It becomes clear that DfAM focuses on the AM process's early phases (I to III). Even though the manufacturing-related phases (IV to VIII) need to be kept in mind during the early design stages, their conduction is not the focus of the DfAM process itself.

FUCHS ET AL. (2022) describe the three main steps of the DfAM process as follows. Similarly to the general PDP (Fig. 2.2) the DfAM process also begins with the *I. Planning and task clarification* phase, in which the specific engineering task is determined, and a list of requirements is derived. This list is the foundation and reference point for the following development process.

After completion of the first phase, phase *II. Conception* follows. Here the functions are determined and structured necessary to meet the requirements. This step is followed by developing initial solution ideas and rough concepts with the support of AM-specific methods, such as design catalogs, creativity techniques, and others. AM design potentials are already considered for finding solutions in the early stages of the DfAM process. This way, the identification of additive manufacturing parts is facilitated, and the purposeful implementation of AM potentials is made possible. With the completion of phase II, the central concept, including functional principles and the product's architecture, is defined.

In phase *III. Embodiment and design detailing*, the computational engineering design and detailing of the 3D part geometries occur. The 3D design is realized with the support of CAD tools in the first place (Sect. 2.1.2). This design phase can be subdivided into *embodiment design*, setting the overall dimensions and partitions of the product, and *design detailing*, where the 3D geometries of the product are finalized. Besides CAD tools, design engineers often rely on additional support from, e.g., simulation and optimization tools (CAE) and AM design rules to generate and finalize optimized 3D part geometries in the AM context. This phase ends with completing the 3D designs of parts, assemblies, components, and the product itself. The subsequent phases are part of the overall AM process, as shown in Fig. 2.29.

Methods, Standards and Norms

In the DfAM context, one main focus is on considering AM design potentials and process-specific restrictions during the ideation and conception phases of the general PDP [Tho+16; Kum18; KWV16]. To do so, DfAM suggests several methods and tools assisting product development ranging from design rules [WW12; Ada15; MWJ17; Kum18], AM specific knowledge databases [BCP12; Wat+19], or additional tools such as the *systematic network of AM design potentials* [Kum18] or the *matrix of conflicting AM potentials* [Fuc+20].

Contemporary work in the DfAM field of research has already gathered and structured available methods and approaches. For example, KUMKE ET AL. (2016) suggest a classification of approaches according to their applicability during the main four phases of the design process. Analyzing Table 2.12, the importance of early consideration of AM potentials in the design process becomes apparent. Many approaches are already partially applicable during *task clarification phase* and mainly focused on the *conceptual design phase*. Whereas, due to manufacturing restrictions, the fulfillment of AM design rules is taken later on during the *embodiment* and *detail design phase* into account (compare [KWV16, p. 9] for the extensive version of Table 2.12).

Additionally to the DfAM literature, standards and norms have been published addressing design challenges in the AM context. Based on the norm VDI guideline VDI 3405 *Additive manufacturing processes, rapid manufacturing—Basics, definitions, processes* [VDI14], that clarifies AM fundamentals, following extensions addressing design-related topics are known (Table 2.13).

Adding to the basics of DfAM described in the subsections above, the following subsections 2.3.3 and 2.3.4 outline AM design potentials and restrictions in more detail, as these are fundamental for the upcoming chapters of this thesis.

Table 2.12 DfAM approaches and methods classified into the four main phases of the design process based on [KWV16]. (● = applicable; ○ = partially applicable; compare also Table 2.3 and Fig. 2.5)

		Task Clarification	Conceptual Design	Embodiment Design	Detail Design
	Phase	I	II	III	IV
AM design rules	Thomas (2009)			●	●
	Seepersad et al. (2012)			●	●
	Adam and Zimmer (2014)			●	●
Application of AM potentials	Bin Maidin et al. (2012)	○	●		
	Watschke et al. (2019)	○	●		
	Schumacher et al. (2019)	○	●	●	
	Kumke et al. (2018)	○	●	●	
	Becker et al. (2005)		●	○	
	Doubrovski et al. (2012)	○	●		
Combined approaches and methods	Ponche et al. (2014)			●	●
	Emmelmann et al. (2011)	○		●	
	Rosen (2007)			●	○
	Leary et al. (2014)			●	

Table 2.13 DfAM-related norms and standards

ID	Title	Source
VDI 3405-3.2	Design rules - Test artifacts and test features for limiting geometric elements	[VDI19a]
VDI 3405-3.4	Design rules for part production using material extrusion processes	[VDI21a]
VDI 3405-3.5	Design rules for part production using electron beam melting	[VDI18]
ASME Y14.46	Product Definition for Additive Manufacturing	[ASM17]

2.3.3 Design Potentials and Related Value Propositions (Customer Benefits)

As summarized in Table 2.10, many product- and process-related benefits are attributed to AM. These benefits result from diverse AM design principles, and engineers face new challenges in purposefully applying them. Thus, several scientific contributions elaborated approaches to structure and systematize this topic (compare *Application of AM potentials* in Table 2.12). The suggested methods and means by the literature have, first of all, one objective in common. To motivate purposeful solution space expansion utilizing additional design freedom and increased manufacturing process flexibility to drive product performance forward and enable novel or not yet feasible business models.

The following structure is suggested in the literature to facilitate the consideration of the numerous AM design potentials in the early development phase in a holistic manner. GIBSON ET AL. (2010) summarizes the unique capabilities of AM with the four complexity categories *shape*, *hierarchical*, *functional*, and *material complexity* [GRS10, 289 ff.].

Definition 21: *Shape Complexity* – One major limitation for part shape complexity in machining is tool accessibility and injection molding, e.g., the necessity for mold separation. The layer-by-layer or voxel-by-voxel characteristic of AM technologies diminishes those limitations as the entire cross-section of each part layer is accessible during the process and can be influenced by, e.g., lasers, light projectors, or nozzles during the build process. This principle enables sophisticated free-form surfaces, hollow space geometries, and features of different sizes and arrangements, enhancing parts' performance. An additional aspect of *shape complexity* also refers to the multiplicity and variability of shapes due to the toolless manufacturing process turning small lot sizes, and individualized shapes economically feasible [Gib+21, p. 560; Kum18, p. 77]. Therefore, *shape complexity*, also known as *form complexity*, summarizes all AM design principles that affect the monolithic part shape on a macroscopic level.

Definition 22: *Hierarchical Complexity* – Similar to shape complexity, AM enables *hierarchical* design across several orders of magnitude in the length scale of a part and specifically in different regions of a part. This complexity refers to the potential of influencing sub-levels of macroscopic part-scale, including mesoscopic and nano-/microscopic purpose-build structures to improve the parts functionality or aesthetics[Gib+21, p. 561]. Here, mesoscopic structures stand for, e.g., surface textures, lattices, honeycombs, or foams [Kum18, p. 79]. Micro- and nanoscopic stand for altering the material structure during the manufacturing process, often causing heterogeneous material behavior. In the LPBF, SLS, or SLA processes, e.g., changes in the laser intensity alter the porosity, structure type, and, thus, the mechanical and thermal properties of the material. In conclusion, *hierarchical complexity* summarizes all design principles that influence the composition of the part on a mesoscopic, microscopic, or nanoscopic level without altering its macroscopic monolithic embodiment.

Definition 23: *Material Complexity* – Additionally to the manipulation of the material structure itself described in *hierarchical complexity*, the category *material complexity* takes the AM potential into account to compose a part out of multiple materials for functional or aesthetic purposes. Depending on the AM process, in addition to varying material regions, graded material transitions in a single part are manufacturable (e.g., DED). Further, in composites, AM-specific potentials such as, e.g., the in situ load-oriented deposition of fibers into a polymer matrix in the FDM process are subordinated to this complexity category [Gib+21, 565 ff.].

Definition 24: *Functional Complexity* – Besides the general potential to reduce partitions in assemblies (often referred to as part consolidation) *functional complexity* also implies design principles that enable, e.g., integrated joints, interlaced and convoluted parts or even embedded components in the parts interior [Gib+21, 563 ff.]. Hence, *functional complexity* results from the possibilities mentioned in the complexities above and summarizes all AM design principles that lead to an expansion of the functional scope of a part. This term definition equates the definition of the term *functional integration* known from the DM context as defined by ZIEBART (2012) [Zie12, p. 112]

Table 2.14 shows exemplary applications that use those complexity categories to achieve a specific product-related benefit. Further, BALDINGER ET AL. (2013) introduce a definition approach of AM potentials based on the interplay of *levers* and *value propositions*. Moreover, AM-related design features are defined as means (lever) to achieve a certain benefit (value proposition) linked to the end customer. Based on the fundamental properties of AM-technologies, six main levers, three primary value propositions, and their interrelations are derived [BLR13, 11 ff.]. Broader, elaborating on this approach KUMKE ET AL. (2018) developed a *semantic network of AM design potentials* condensing this manifold topic of AM potentials

Table 2.14 Overview of DfAM-related design potential categories derived from [GRS10, p. 289] including exemplary applications. (*shot at Formnext exhibition - Frankfurt am Main*)

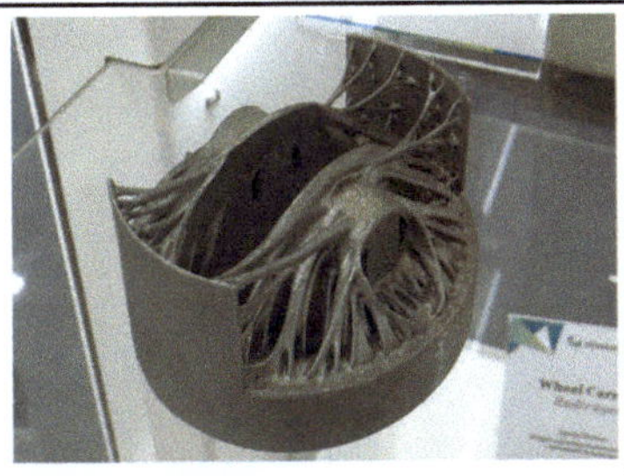

Category **Shape Complexity**
Description Topology optimized piston
Exhibitor Hexagon (2019)
Benefit Improved stress peaks and cooling properties
Status Prototype

Hierarchical Complexity
Exhaust finisher with lattice structures
AM Power, Bugatti Engineering GmbH (2019)
Reduced heat conduction
Functional prototype

Category **Material Complexity**
Description Multi-material after burner
Exhibitor SLM Solutions, Fraunhofer IGCV (2022)
Benefit Improved heat transfer
Status Prototype

Functional Complexity
Drive shaft with gears, cooling, lubrication
DMRC (2019)
Part consolidation, design space reduction
Prototype

holistically and comprehensively [Kum+18]. Here the levers are structured based on the four complexity categories (Table 2.14) introduced by GIBSON ET AL. (2010) and further broken down into numerous engineering design principles (Fig. 2.31). The value propositions pursued by the application of those levers are subdivided into three main clusters *additional product value*, *cost reduction*, and *indirect value propositions*. Also, an *improvement in sustainability* and *reduced time to product/market* are pointed out as potential outcomes. Regarding the complexity of the interrelations, an interactive version of the semantic network is necessary for the productive application of the method. With this approach, product developers can quickly access DfAM-specific knowledge to solve a particular design challenge purposefully. Therefore, this systematic understanding of AM potentials as the combination of levers and value propositions, including all AM design principles, is fundamental for this work.

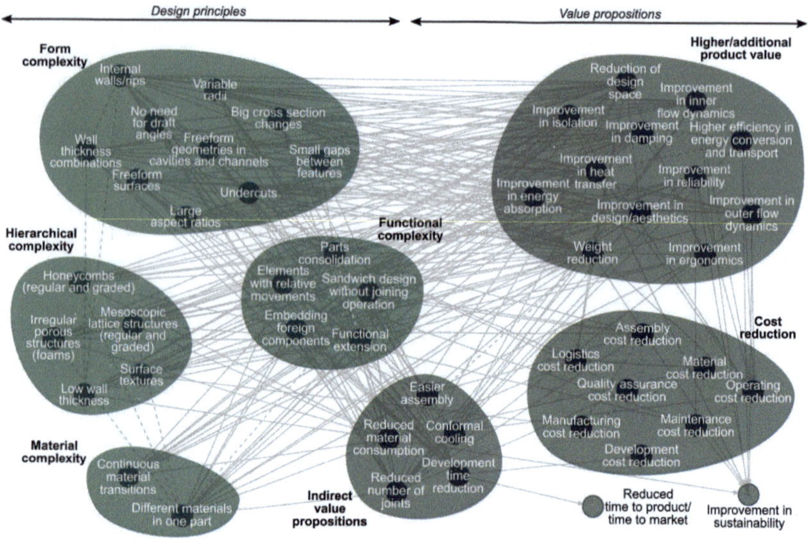

Fig. 2.31 The semantic network of AM design potentials distinguishing design principles and related value propositions based on [Kum+18, p. 485]

2.3.4 Process Restrictions and Resulting Design Rules

Supplementary to the motivation and propagation of AM-specific design freedom, the DfAM research field also addresses manufacturing restrictions in order to apply AM potentials reasonably [HMS04; AZ14; AZ15; Gao+15; Tho+16; Kum18]. Commencing with design restrictions such as the consideration of support structures over to heterogeneous material properties and layer dependent surface qualities as well as post-processing efforts and productivity limits, in DfAM process specific constraints are examined extensively. In this context, it becomes clear that process-specific design knowledge positively influences AM part properties and

Table 2.15 General design rules in the context of DfAM free translated from [Kum18, p. 114] extended with exemplary samples

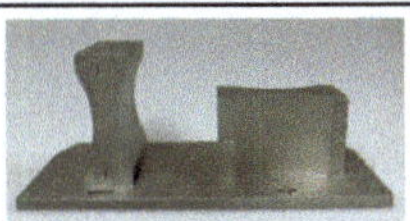

1. Dimensional accuracy:
Accuracy and resolution is limited, e.g., by the layer thickness or the layer generating component (e.g., a laser beam). Typical layer thicknesses range from 0,03-0,2 mm.

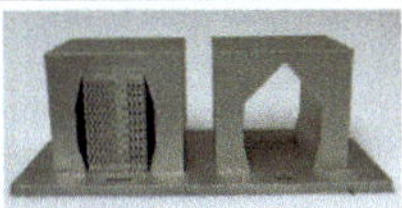

2. Support structures:
Some processes rely on support structures for stabilization. The amount of support structures needed is related to the part build orientation. These are particularly necessary for overhangs below certain angles between the part and building platform. Removal is almost always during post-processing needed.

3. Rough Surfaces:
The surface quality is often relatively rough. The part's orientation in the build chamber mostly influences it.

4. Anisotropic Material:
Material properties might be anisotropic and related to the orientation of the part in the build chamber. Therefore, the part's mechanical properties are often lower when stressed perpendicular to the building platform.

5. Powder removal:
Powder bed-based processes require powder removal openings in the part to clear voids and internal ducts in post-processing

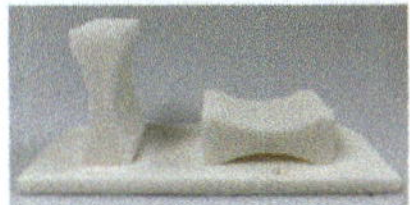

6. Limited part size:
The maximum part size is determined by the build chamber size and might require partitioning

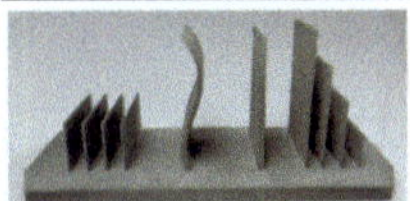

7. Minimum feature size:
Minimum dimensions apply for geometrical features such as wall thicknesses

8. Minimum spacing:
If e.g., minimum spacing is considered, interlaced/convoluted movable parts can be realized in some processes

9. Choice of material:
Material selection is limited to the specific process. Integrating multiple materials in one part is only possible in some processes.

10. Post-process:
When post-processes, e.g., machining are required for part finishing, their specific restrictions need to be considered (e.g., tooling accessibility).

pre- and post-processing efforts. Table 2.10 summarizes some AM restrictions as product-related and process-related challenges. One approach to comply with process-specific restrictions is to consider them during the design process (DfM—Design for Manufacturing Sect. 2.1.1). This consideration of restrictions can be done by following appropriate design rules as suggested in the DfAM strict sense methods [KWV16]. Those rules are often structured and gathered in design catalogs [Tho09; See+12; AZ14; Kum18]. KUMKE (2018) elaborated on this topic and introduced a checklist with the ten most important AM design rules from his perspective (Table 2.15).

If design rules such as those mentioned in Table 2.15 are not considered in the early stages of the design process, a false assessment of the feasibility of the part with AM is possible, the resulting part quality/performance can be sub-optimal, and unnecessary efforts during port-processing can be caused. Exemplary in more detail, support structures are necessary to build part overhangs, internal channels, or cavities, depending on the process of choice. Those structures are prone to tedious removal during the post-process. The avoidance of specific angles in the design (e.g., the second design rule *support structures* in Table 2.15) or by a wisely chosen build orientation the need for support structures can be mitigated beforehand [Ada15; GH16; Kum18; Wat+19; Fuc+22]. Even though being of fundamental value for AM applications, those design rules are specific to each AM process and are therefore only generally considered in the elaborations in this thesis.

2.4   Implications for this Work

Based on the fundamentals introduced in the previous sections, this section summarizes the limitations of each topic and their implications for this work.

2.4.1   Summary and Challenges in Design Methodology

The fundamentals section begins by giving insights into Design Methodology (DM) in the context of industrial product development (PD) see Sec. 2.1. Product Development Process (PDP), Design, Design Theory (DT) and Methodology (DM, DTM), Systems Engineering (SE), as well as Design for X (DfX) are relevant terms for the scope of this work. DM is relevant for the beginning of the Product Life Cycle (PLC) and focuses mainly on the following phases of the PDP: Task Clarification, Conceptual Design, and Embodiment Design. DM is crucial to cope with increased development complexity, yet DM is constantly in a dilemma between general validity and applicability to specific use cases (Sect. 2.1.1). The literature suggests numerous general and specific design methods and approaches, some of which have been established to standards and norms over the years (Sect. 2.1.1). Computer-aided engineering (CAE) and computer-aided design (CAD) tools are essential in DM. They are increasingly interwoven, covering each phase of the PDP in the sum of its numerous manifestations (Sect. 2.1.2). Virtual geometry representation techniques are a core element of CAx, and several approaches with varying suitability can be distinguished from one another. The parametric associative CAD workflow is most common in technical applications, enabling design engineers to define product assemblies in virtual 3D space. As a result of the continuous Mass Individualization (MI) trend in modern society, product variants are drivers of corporate complexity, particularly during the PDP (Sect. 2.1.3). Increased product, process, and manufacturing flexibility challenge design engineers and demand next-level computer-aided specification tools to cope with variant-related product complexity.

Function-oriented system engineering design, inter-disciplinarity, and multi-objective optimization can be considered overall challenges for advances in DM. Contemporary research is focused on the evolutionary progress, combination of established design methods, and specialization of approaches considering those challenges. Trade-offs such as manual labor-intense design workflows versus cost-cutting in PD, and decreasing corporate complexity through variant reduction versus mass individualization trends in society are additional fields of tension, demanding complexity management and innovations in engineering methods, processes, tools, and manufacturing technologies. In addition, continuous advances in computer-aided optimization techniques based on *Finite Element Method* (FEM) pave the way for new design DM approaches towards multi-disciplinary optimization and multi-physics design. These challenges are crucial to the context of this work. With additional advances in those aspects, complex optimized designs that take advantage of DfAM design freedom are enabled to maximize product-related benefits achieved with AM. Therefore, in the following chapters, specific automotive applications addressing those engineering design challenges are elaborated in further detail.

2.4.2 Summary and Challenges in Vehicle Conception

After industrial PD in general, fundamentals of vehicle conception (VC) are clarified in Sec. 2.2. Besides the essential interdependencies in vehicle physics (Sect. 2.2.1), the most governing boundary conditions of VC, also principles of the composition (Sect. 2.2.2) and conception process of a vehicle (Sect. 2.2.3) are explained.

Thereby, the sensitivity of physical vehicle characteristics is arbitrated to enable the evaluation of respective vehicle performance improvements. Further, a shared perception of the vehicle as a holistic system consisting of separately developed yet synchronized technical sections is essential for the following elaborations of this work. Major vehicle-defining decisions based on technical feasibility are made during the conception process, narrowing down the solution space available for the following development phase. Finding the right compromise between mutually dependent requirements and properties is crucial for aligning customer-relevant vehicle characteristics with the target group of customers and, therefore, for customer satisfaction. Sect. 2.2.4 emphasizes that those conflicting objectives are a significant challenge during engineering design in VC, constantly demanding new degrees of design freedom enabled by innovative technologies.

Current challenges in the automotive industry that need to be considered during VC can be categorized as product-related and process-related. The first crucial product-related challenge is that the drive-train transition from internal combustion engines to battery electric vehicles significantly impacts VC. Besides the impact of this transition on the traffic infrastructure, particularly regarding the vehicles' technical package, battery electric drive train is unlocking additional space in the vehicle's interior without lengthening external dimensions. Simultaneously the drawback of an increasing overall curb weight needs to be considered, e.g., in driving and crash behavior. Due to the still costly drive-train technology, the battery, in particular, scale effects are key to decreasing overall retail price. This circumstance, combined with advanced autonomous driving functionalities, induces increased demand towards modularity and digitization in upcoming vehicle platforms challenging VC anew.

Process-related challenges are derived from the product-related aspects mentioned above. The number of functions implemented in current vehicles uncovers new complexity levels in their functional architecture, mainly due to digital or software-related functions. The ongoing paradigm shift of automotive OEMs transitioning from solely vehicle manufacturers to mobility providers (e.g., shared mobility) and the resulting new lines of business [SMT11, p. 14; Kal+22, p. 80] present additional system-driven requirements on upcoming vehicles to be considered during VC (e.g., provision of real-time vehicle data to third parties such as fleet operators, meteorological services, and road maintenance depots). To manage this, complex development processes are being reoriented towards systems engineering and function-oriented development, further establishing development methods such as the V-Model in the industry. Lastly, challenges derived from the multi-disciplinarity of engineering tasks in highly disciplinary development structures must be mentioned in this context. Each technical section of a vehicle is in itself highly specialized and optimized. The coordination and optimization of their interdependencies is yet a challenge.

Nevertheless, besides PD, DM, and automotive engineering in general, there needs to be more contemporary literature addressing findings focused on VC processes and methods. VC as an engineering discipline is complex, highly methodological, company and culture-specific, and deeply integrated into the core business so that comprehensive research is impeded. As mentioned in the following subsection, AM can distort the available solution space for design engineers, recalibrating what is technically feasible. Therefore, it must be considered during the product conception phase and, thus, during VC. The following chapters elaborate on this distorted design solution space due to the consideration of DfAM in the automotive context. Novel engineering design approaches are analyzed, and encountered limitations are further specified.

2.4.3 Summary and Challenges in Design for Additive Manufacturing

On the one hand, the basics of AM are summarized in the fundamentals Sect. 2.3 Design for Additive Manufacturing—Potentials and Restrictions. Here, besides term definitions, origins, and general placement of AM in the manufacturing process classification, exemplary applications in the automotive context are outlined. Further, the general AM process is explained in eight steps, whereas the main steps from *IV Data Preparation* to *VIII Part Finishing* are clarified in more detail (Fig. 2.29). Fig. 2.30 overviews the numerous process types assigned to AM. Furthermore, the technology's product- and process-related benefits and restrictions are structured in Table 2.10.

On the other hand, the field of research DfAM is outlined by introducing the AM process steps in focus and appropriate methods, standards, and norms pointed out. Design potentials are structured into the four complexity types (*shape, hierarchical, material,* and *functional complexity*). Additionally, contemporary work on the systematic of AM potentials composed of the combination of design principles and value propositions is pointed out. Also, an awareness of AM process-specific restrictions and the consideration of resulting design rules in the early design phase is raised (Sect. 2.3.4).

Besides product-related limitations such as surface quality, material properties, and dimensional accuracy, AM technologies face process-related challenges such as reproducibility, scalability, economic efficiency, material certification, and others (Table 2.10). Those aspects need to be improved to drive the implementation of the technology in the industrial context forward.

Nevertheless, the state-of-research also uncovers shortcomings in the DfAM context that limit the implementation of AM. A focus point of current challenges is identifying purposeful applications in an industrial small to medium-lot-size setting. In the automotive industry, e.g., applications are often either not yet profitable or need to leverage additional benefits for customers and companies that cannot be leveraged conventionally (value propositions) to justify AM. One common approach in this context is *part substitution*. Here, a particular part of a vehicle (product) is slightly modified according to basic DfAM rules, 3D printed, and put back into its initial place in the assembly, not altering its original set of functions. In doing so, the beneficial impact of such an optimized part for the entire product as a system might be relatively low and, in many cases, does not sufficiently substantiate additional manufacturing efforts. Therefore, applications following this approach often do not withstand the assessment phase. So, a lack of AM applications and approaches is identified that question conventional component partitions and their functional scopes while holistically taking advantage of AM potentials, focusing on additional product or process value.

However, some state-of-the-art cases already aim to apply AM potentials on complete vehicle level. Even though those cases are discussed in more detail in Sect. 3.1, it becomes clear that those applications still need to rely on scientific methods to leverage AM potentials holistically. Thus, a lack of methods addressing the scalability of AM potentials up to complex products such as automotive vehicles is identified. Additionally, particularly in the complete vehicle level context, a strong correlation between limitations in engineering methods and engineering tools (Sect. 2.4.1) is identified to enable multi-disciplinary optimized AM applications. A holistic analysis of automotive AM applications based on systematic and scientific DfAM methods is necessary to research this technology's potential impact on vehicle design further. This analysis begins in Chap. 3.

Concluding, as mentioned in Sect.1.2, in order to dissociate from traditional engineering design patterns, propel the imagination of engineers, and enhance the capabilities of their tools, it becomes clear that a paradigm change in engineering design is necessary to take advantage of design freedom and product complexity enabled by AM systematically. Further, an increased development complexity presents an engineering challenge in this context. This correlation implies necessary alterations of the hands-on engineering design process considering interferences between potentially opposed AM design principles.

The elaborations in the following chapters of this work contribute to the combined limitations described in the three above-mentioned topics DM, VC, and DfAM.

Potential Benefits of Additive Manufacturing in the Vehicle Conception Context

3

A complete vehicle represents a complex technical system highly optimized for its targeted customer group. It consists of mainly four technical sections (*vehicle body, chassis, drive train, and electric/electronics*) describing the first layer of its functional architecture (Sect. 2.2). Those sections can be comprehended as subsystems or disciplines in which development activities are organized. Additive manufacturing (AM) potentials, particularly considering design principles towards functional integration and part consolidation, indicate a beneficial impact to a technical system superior to its part-level. Therefore, this chapter aims to combine these contexts by systematically linking AM design potentials to major vehicle properties. This way, AM application areas independent of part-level and conventional part partitions can be identified that might significantly impact vehicles' customer-oriented performance.

First, state-of-the-art applications are gathered and analyzed in Sect. 3.1. Different approaches to this topic of interest can be synthesized. In a second step in Sect. 3.2, potential AM application categories on the whole vehicle level are derived using an abstraction approach to the vehicle's main characteristics. Nevertheless, before further elaborating on those application categories in Chap. 4, not yet identified mutual dependencies between AM design principles are addressed in Sect. 3.3. This enables a differentiated consideration of AM potentials for vehicle conception (VC) in the following. This chapter is summarized with its findings in Sect. 3.4.

Supplementary Information The online version contains supplementary material available at https://doi.org/10.1007/978-3-658-49677-7_3.

D. Fuchs, *Exploration of the Influence of Additive Manufacturing Design Potentials on Vehicle Conception*, AutoUni – Schriftenreihe 179,
https://doi.org/10.1007/978-3-658-49677-7_3

3.1   State-of-the-Art of Additively Manufactured Vehicles

Beyond the automotive applications of AM shown in Table 2.11 in Sect. 2.3.1, the state-of-the-art use cases mentioned in this section are focused on applications in which the AM technology plays a relevant role on overall vehicle level. This perimeter means that either main vehicle components are additively manufactured, or the vehicle concept is altered and focused on AM characteristics. Even though prototyping is an important application area, prototype applications in which major vehicle components are only additively manufactured to embody early demonstrators of later conventionally manufactured vehicles are also not the focus of this research. Nonetheless, different approaches and showcases to this topic are described in more detail. The varying technological readiness level of the showcases mentioned in this section is a crucial indicator for the authenticity of the respective approach and, thus, fundamental for an assessment.

Table 3.1 overviews an exemplary range of vehicles realized with AM as significant manufacturing technology. The listings are ordered from the top left to the bottom right corner according to the increasing AM penetration level in the vehicle concept and its respective requirement aspiration. The examples listed range from non-drivable show cars for demonstration purposes to drivable functional pre-production models.

Analyzing the examples listed in Table 3.1 it becomes clear that different value propositions motivate the application of AM as significant manufacturing technology during VC. Those value propositions can range from production costs reduction of small lot sizes (individualization), over to production on location, up to lightweight and bionic design. Similarly, different are also the areas of operation and, therefore, the targeted customer groups for those vehicles ranging from people movers with low volume-/range-/speed-requirements up to high-performance sports cars.

Further, Table 3.1 shows that no examples have yet reached the development status *series/production vehicle*. The examples are either under ongoing development or discontinued without reaching production status. Also, a concept that targets the fulfillment of high requirements and lot sizes is not represented. This context reflects the primary product- and process-related AM challenges mentioned in Table 2.10, e.g., reproducibility, scalability, and economic efficiency. The resulting approach types and the derived application areas for AM vehicles are explained next.

Unique Show Cars (One-Offs)

Based on the characteristics of most AM processes to enable cost-efficient small lot sizes, one application area of AM in VC is show cars. These one-off concept

Table 3.1 Exemplary overview of additively manufactured vehicles. (Note: *This listing is not exhaustive, the information gathered is in parts subject to the author's interpretation*; legend: *N = name; M = manufacturer; DS = development status; D = driveability; AM = AM scope; P = AM process; V = Value proposition; S = Source; Y = year*)

N(M) **Urbee** (KOR Ecologic) - *discontinued*
DS(D) Functional demonstrator (drivable)
AM(P) Body panels (FDM)
    V Production cost reduction of small-lot-sizes
    S(Y) [Str18](2013)

Venture VIII (Nanyang Techn. Univ. Singapore)
Research vehicle (drivable)
Load bearing body panels (FDM, SLS, PJM)
Lightweight design, Individualization
[San15](2015)

N(M) **Strati** (Local Motors) - *discontinued*
DS(D) Functional Demonstrator (drivable)
AM(P) Monocoque structure (FDM/BAAM)
    V Assembly cost reduction, production on location
    S(Y) [Spe14](2014)

Light Cocoon (EDAG Engineering Group AG)
Show car, technology demonstrator (non-drivable)
Body skeleton (SLA), frame nodes (LPBF)
Lightweight/bionic design
[EDA15](2015)

N(M) **LSEV** (XEV)
DS(D) Pre-production model
AM(P) Body and interior panels (FDM)
    V Production cost reduction of small-lot-sizes
    S(Y) [3Dn18](2018)

Blade I (Divergent Technologies Inc.)
Technology demonstrator (drivable)
Trellis frame nodes (LPBF)
Lightweight design, production on location
[Wei16](2016)

studies embody how future vehicles might look. These are primarily non-drivable models carrying novel ideas and are built for presentation purposes only, e.g., for exhibitions. For the context of this work, those concepts that incorporate design potentials uniquely to AM are in focus.

Table 3.2 gathers a few show car examples with the applied product- or process-related AM potentials. For example, the *Light Cocoon* concept focuses on the lightweight design of the vehicle body structure and exterior taking advantage of the design freedom and advanced surface quality enabled by the SLA process. In this case, a new approach to vehicle design is suggested in which the vehicle structure is understood as a combination of a bionically optimized load-bearing skeleton and a weather-resistant fabric overlay (compare conventional vehicle body types in Sect. 2.2.2 Table 2.9). In the second design concept *Fractal*, other AM design freedom is utilized by implementing a functional structure on the vehicle's interior panels to enhance their sound-absorbing properties. The third example of a *Shelby Cobra* replica takes advantage of the process-related benefits of FDM (in

Table 3.2 Exemplary overview of additively manufactured show cars. (Note: *The information gathered is in parts subject to the author's interpretation;* legend: *B = body; C = chassis; D = drive train; E = electrics/electronics*)

Light Cocoon (EDAG Engineering Group AG, 2015) [Sta15; EDA15]

Show car, technology demonstrator (non-drivable)

Body skeleton (SLA), frame nodes (LPBF)

Lightweight/bionic design

 C D E

Fractal Concept (Peugeot, 2015) [Die15; Peu15]

Show car, (drivable)

Interior panels and parts (SLS)

Improvement in sound absorption and aesthetics

 C D E

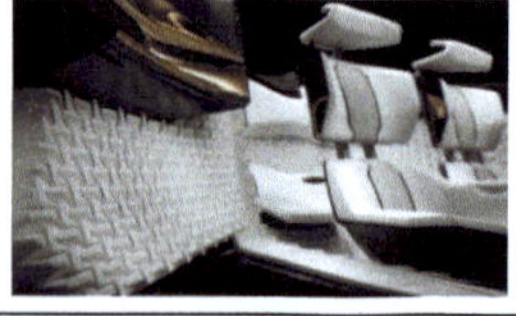 

Shelby Cobra Replica (Oak Ridge National Laboratory, 2015) [ORN15]

Show car, homage (drivable)

Body panels and structure (FDM/BAAM)

Toolless manufacturing

 C D E

this dimension, also known as BAAM), enabling the manufacturing of large rough parts in small lot sizes without relying on specific molds or tools. These three examples accentuate that in show cars, AM applications are mainly focused on the body as one of the four technical sections of the vehicle (compare *technical sections* in Sect. 2.2.2 Fig. 2.20).

Individualization Focused Vehicle Concepts

The following examples are also based on the suitability of AM technologies for small lot sizes, particularly their independence of product-specific tools or molds. However, those characteristics are not just used to realize unique one-off show cars but instead applied to enable unique vehicles for the end customer. As specified in Sect. 2.1.3 Fig. 2.14, there are different approaches in PD to personalize, customize, or individualize a product to reach a customer-specific or customer-specified solution. The examples gathered in Table 3.3 aim to combine the contexts of AM characteristics and PD individualization approaches to offer the customer additional product value.

The concepts *Daihatsu Copen* (compact car) and *Toyota i-Road* (urban commuter), e.g., incorporate additively manufactured (FDM) interchangeable body panels in the front and rear sections of the vehicles. Those panels should be customized to the customer's preference altering the vehicle's exterior appearance with differing 3D surface patterns and colors. With the possibility to individualize additively manufactured body panels, the modular *Honda Micro Commuter* concept follows a similar approach yet targets corporate customers in the first place, e.g., delivery services. In addition to those appearance-focused individualization approaches, the concept *Fiat Centoventi* (compact car) follows a functional focus for additively manufactured vehicle interior components. Basic accessories such as cup holders are supposed to be fitted or even printed by customers. Concluding also in concepts with individualization focus, the vehicle body is the technical section AM potential implementations are focused.

Low Requirement Vehicle Concepts

Another vehicle concept type taking advantage of AM potentials can be categorized as *low requirement vehicles*. These vehicles are purposefully designed for a specific application area with diminished requirements to standard vehicle registration classes for public roads. These concepts are subject to low operating speeds, safety, comfort, and quality demands, among others, and exhibit a simplified layout. These areas of applications represent a lower threshold for AM processes with large build volumes yet limited accuracy and surface quality (e.g., BAAM). Table 3.4 illustrates examples in this context. The former start-up Local Motors, e.g., introduced three

Table 3.3 Exemplary overview of individualization focused vehicle concepts enabled by AM. (Note: *The information gathered is in parts subject to the author's interpretation;* legend: *B = body; C = chassis; D = drive train; E = electrics/electronics*)

Copen (Daihatsu, 2015) [Str22] Concept car (drivable) Body panels front/rear bumper, bonnet (FDM) Product differentiation, customized aesthetics **B** C D E	 → 
i-Road (Toyota, 2013) [Toy15] Functional prototype (drivable) Body bonnet panel (*not known*) Product differentiation, customized aesthetics **B** C D E	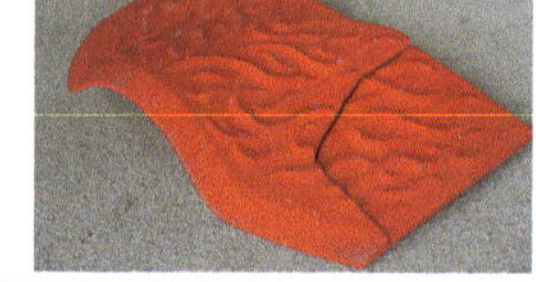 → 
Micro Commuter (Honda, 2016) [Atl16] Concept car (non-drivable) Body panels (FDM, PJM) Product differentiation, customized advertisement **B** C D E	 → 
Centoventi (Fiat, 2019) [Mal19] Concept car (non-drivable) Interior components (*not known*) Product differentiation, customized functionality **B** C D E	 → 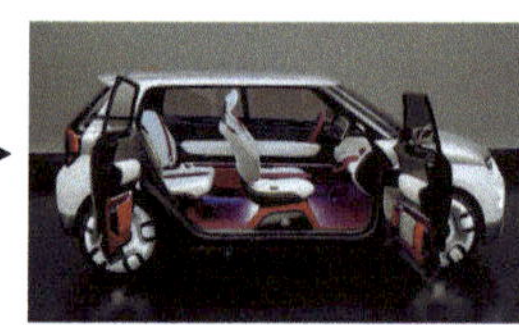

consecutive low requirement vehicle concepts: *Strati* (Table 3.1), *LM3D Swim,* and *Olli* composed of an additively manufactured monocoque design incorporating vehicle structure and body panels simultaneously. A similar approach to vehicle design, yet based on a platform and upper body structural architecture, can be observed in the concepts *Jeep* and the *Shelby Cobra Replica* (Table 3.1) elaborated by the Oak Ridge National Laboratory (ORNL). Those BAAM-based concepts use polymer pellet basis material reinforced with short-fibred carbon conventionally used for injection molding [Cha+17, p. 101].

Table 3.4 Exemplary overview of low requirement vehicle concepts enabled by AM. (Note: *The information gathered is in parts subject to the author's interpretation;* legend: *B = body; C = chassis; D = drive train; E = electrics/electronics*)

LM3D Swim (Local Motors - *discontinued*, 2015) [Fun15]

Functional prototype (drivable)

Body structure, exterior and interior panels (FDM/BAAM)

Assembly and logistics cost reduction

B C D E

Olli (Local Motors - *discontinued*, 2016) [Cla17a; Cla17b]

Functional prototype (drivable)

Platform exterior and interior panels (FDM/BAAM)

Assembly and logistics cost reduction

B C D E

Jeep (Oak Ridge National Laboratory, 2015) [Cha+17]

Research vehicle / technology demonstrator (drivable)

Platform and upper body structure (FDM/BAAM)

Assembly and logistics cost reduction

B C D E

High Performance Concepts

Besides low-requirement applications, also applications in high-performance sports cars are already known. Here a particular focus is on lightweight design through shape and hierarchical complexity. Some part-level examples of this kind are shown in Sect. 2.3.1 Tab 2.11 and Sect. 2.3.3 Table 2.14. Three differing approaches on how this principle can be applied on complete vehicle body are shown in Table 3.5.

The vehicle concept *Blade*, introduced by Divergent Technologies (also known as Czinger Vehicles), started in its first development status with a trellis frame approach. This frame consisted of carbon tubes and additively manufactured (LPBF) aluminum nodes. The body panels and chassis components are manufactured separately and do not show a particular lightweight focus. According to the second development status of the concept *Blade II* already displays a more integrated vehicle architecture intermediate to a space frame or monocoque design, with widened profile cross-sections

Table 3.5 Exemplary overview of high performance vehicle concepts enabled by AM. (Note: *The information gathered is in parts subject to the author's interpretation;* legend: *B = body; C = chassis; D = drive train; E = electrics/electronics*)

Blade I (Divergent 3D, 2015) [Wei15]

Prototype (non-drivable)

Body structure nodes (LPBF)

Weight reduction

B C D E

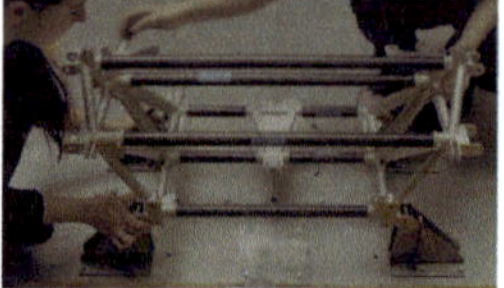 

Blade II (Divergent 3D, 2016) [Wei16]

Functional prototype (drivable)

Body structure and chassis components (LPBF)

Production on location, production cost reduction of small lot sizes, weight reduction

B C D E

Czinger 21C (Divergent 3D, 2022) [Eva22]

Pre-series vehicle (drivable)

Body structure, chassis, and drive train components (LPBF)

Production on location, production cost reduction of small lot sizes, weight reduction

B C D E

of the aluminum nodes and added curved carbon shell components. Also, chassis components, e.g., wishbones and wheel carriers, are additively manufactured. The third iteration of the concept, *Blade III*, has been developed further towards unibody modular shell design, considering additional assembly-process-specific aspects. In addition to body and chassis components, drive train components such as exhaust pipes and hear shields are additively manufactured. Thus, the *Blade* concept shows the highest penetration level of AM in a vehicle concept yet.

Generative Concepts

The last distinguishable type of approach in the state-of-the-art can be observed with the *La Bandita* concept from the start-up *Hackrod* (*discontinued*) shown in Table 3.6. This concept focuses on the digital characteristics and the resulting automation potential of AM processes and aims to combine generative or algorithmic design methods with AM. The herewith intended automation of the design process is crucial to enable self-individualized vehicles by the end customer, also known as *Co-Creation* (compare Fig. 2.14). In this concept, sensor-captured data is used to feedback on the vehicle's and the driver's physical factors. Based on this data, the vehicle body structure is generated in a parameterized system [ZW20, p. 2; Tra21, 88 ff.].

Table 3.6 Exemplary overview of a generatively designed vehicle concept enabled by AM. (Note: *The information gathered is in parts subject to the author's interpretation;* legend: *B = body; C = chassis; D = drive train; E = electrics/electronics*)

La Bandita (Hackrod - *discontinued*, 2016) [Sco16; Glu18]

Prototype (drivable mock-up)

Body trellis structure (DED) and panels (FDM/BAAM)

Production on location, production cost reduction of small lot sizes, weight reduction

B C D E

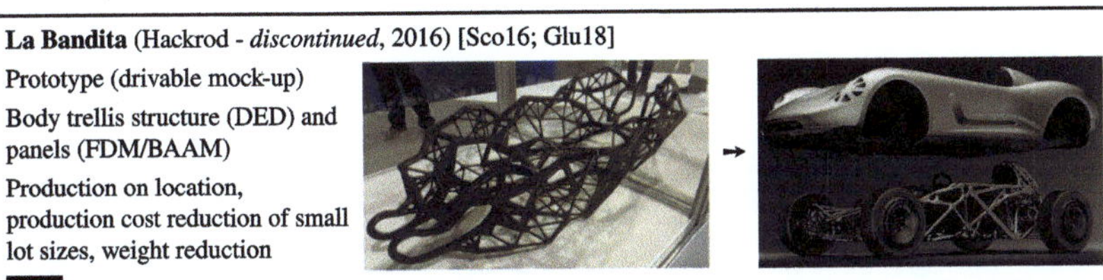

Conclusion and Comparison with Conventional Body Design Approaches

In conclusion, five basic types regarding the differing approaches in the state-of-the-art AM applications on the complete vehicle level mentioned above can be summarized. These approaches target to optimize different vehicle concept characteristics as described in Fig. 3.1. These concepts originated from a practical environment and aim to display what is manufacturable with AM technologies. None of those approaches passed the development milestone start of production (SOP) yet and they do not necessarily represent feasible or scientific approaches to vehicle design developed by specialists for automotive mass production. Also, approaches that aim to leverage DfAM potentials holistically are not yet known. The concepts mentioned above emphasize a particular value proposition of AM not yet taking advantage of the full AM potential for vehicle conception. Therefore, scientific and methodological support is necessary in this context.

Focusing on the vehicle's body as the most addressed technical section for AM applications on the whole vehicle level, it becomes clear that several of the conventional vehicle body design methods introduced in Table 2.9 in Sect. 2.2.2 are being addressed with the AM based vehicle concepts. A correlation matrix shown in Fig. 3.2 clarifies relations and similarities between conventional and AM-based vehicle body design approaches, thereby distinguishing between polymer and metal-based AM concepts.

Fig. 3.1 State-of-the-art overview of AM application approaches on vehicle conception level

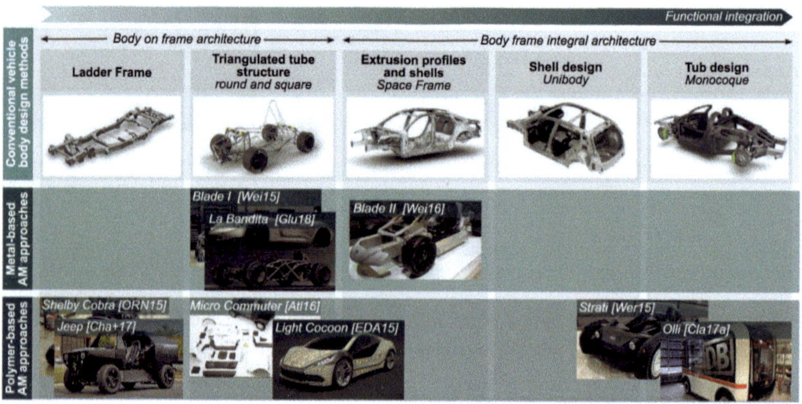

Fig. 3.2 Comparison of conventional and AM-based body design approaches in the state-of-the-art and resulting white spots

Besides the similarities, Fig. 3.2 also points out potential white spots in which no AM approach is known to the state-of-the-art. For example, an approach has yet to follow a metal-based body shell or monocoque design. Also, a space frame design based on a polymer AM process is unknown. Additionally, most concepts follow a separate body and frame architecture, only a few aim for an integrated approach. These aspects are discussed in more detail in Sect. 5.2.1.

In the following AM potential categories for vehicle conception are systematically derived to enable a holistic approach to this context.

3.2   Categorization of AM Potentials for Vehicle Conception

As analyzed in Sect. 3.1, there are no approaches known in the literature in which DM and DfAM methods are applied simultaneously and systematically to face challenges in vehicle conception with AM design freedom. With this objective in mind, this section focuses on identifying significant fields of action for a purposeful application of AM potentials on a complex system such as a vehicle. According to PISCHINGER AND SEIFFERT (2016), there are several reasons for the development or the implementation of new technologies in the automotive context [PS16, p. 50]:

- Improvement of present properties
- Achievement of so far unfeasible properties
- Cope with conflicting objectives
- Sustainability improvements of vehicle and traffic
- Increase in economic efficiency
- Realization of manufacturer's strategic goals

As customer needs determine product demand and thus the product's economic success, a customer-oriented vehicle concept alignment is crucial for successful product development. In contrast, novel technologies enable product innovations, which in the sum of their properties, might surpass the customer's expectations yet be of substantial competitive importance. Therefore, successful product development needs to be receptive to customer needs (*market pull*) as well as make room for innovations (*technology push*). Considering that the focus of this work is to analyze the impact of AM potentials on the conception of a vehicle, the following elaborations set emphasis on product- rather than on process-related innovations (Table 2.10). Even though, remember that in the AM context, both aspects can be mutually dependent. Thus, the first four reasons for implementing new technologies mentioned above are prioritized as they address the product's properties in the first place. As a consequence of this relation, the chosen approach for the following elaborations in this subsection targets *conflicts of objectives* as the main limitation in VC (Sect. 2.2.4, 2.4.2). This approach is based on the hypothesis that by applying AM potentials to cope with conflicting objectives in VC, customer-relevant value propositions are constantly in focus.

First, fundamental generalized vehicle properties are derived, holistically summarizing the complex multitude of vehicle requirements (Sect. 3.2.1). Second, in Sect. 3.2.2, based on an idealization of those fundamental vehicle properties, discrepancies between a conventional and a fictitious optimized set of vehicle properties are identified, resulting in significant conflicting objectives in VC. Third, AM potentials are systematically applied to optimize the conflicting objectives (Sect. 3.2.3). Here, DfAM-enabled design concepts for potential innovations that should be considered during VC are generated.

3.2.1 Derivation of Fundamental Vehicle Properties

The derivation of fundamental vehicle properties in this subsection is based on the following three different sets of vehicle requirement collectives:

- Network of automotive requirement categories [PS16]
- Parametric vehicle model [Pri11]
- Exemplary vehicle property structure derived from a passenger car (Volkswagen AG)

Whereas the first set of requirements only includes the major categories, the second and third sets endeavor to cover the vehicle holistically.

PRINZ (2011) developed an approach to describe a vehicle holistically with a parametric relation system of 98 parameters [Pri11, 60 ff.]. This complex set also includes the parameters' interdependencies and aims to facilitate the generation of vehicle concept variants. Fig. 3.3 summarizes the basic structure of this model consisting of six relation system categories.

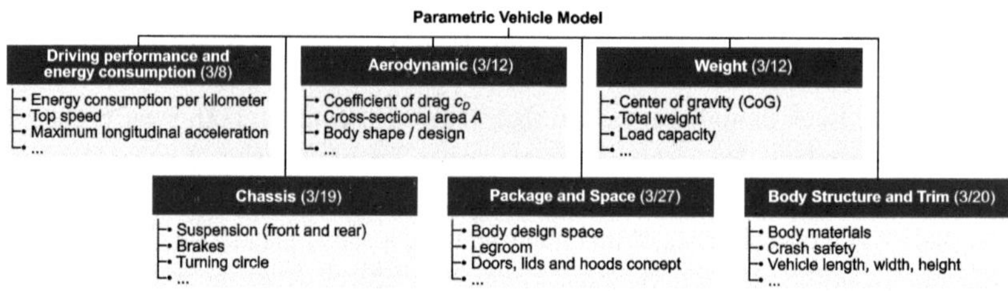

Fig. 3.3 Main categories of a parametric relation system of a holistic vehicle model based on [Pri11]

The parameters gathered in this model are unambiguous, quantifiable, and thus, in their entirety, necessary to define a specific vehicle from a technical perspective. Therefore, those attributes are not directly comparable with the vehicle requirement categories described in the literature (Sect. 2.2.3). Here, PISCHINGER AND SEIFFERT (2016) relate, e.g., the category *comfort* with factors addressing, among others, driving, vibration, acoustic, and ergonomic comfort. Whereas, according to PRINZ (2011) comfort-related parameters can be identified in several categories, e.g., in the *chassis* (*suspension comfort*), or the *package and space* (ergonomic comfort) category. Concluding, the parametric vehicle model from PRINZ (2011) is focused on coherences of the vehicle physics context (Sect. 2.2.1).

The vehicle development process in an international OEM is a highly systematized routine, enabling rapid development of novel models (Sect. 2.1). A different approach is used to guarantee the fulfillment of all requirements concerning the vehicle concept to be developed. Here, a *catalog of requirements* is used in the early stages of the process, in which the vehicle-defining requirements are compiled. This catalog specifies the entire vehicle's property structure and can consist of, e.g., eleven main requirements categories, summing up to more than 800 specification entries. Therefore, a holistic evaluation of such a collective body of requirements is challenging. Nevertheless, Fig. 3.4 illustrates the main generalized requirement categories of such a catalog that are particularly relevant for an automotive OEM.

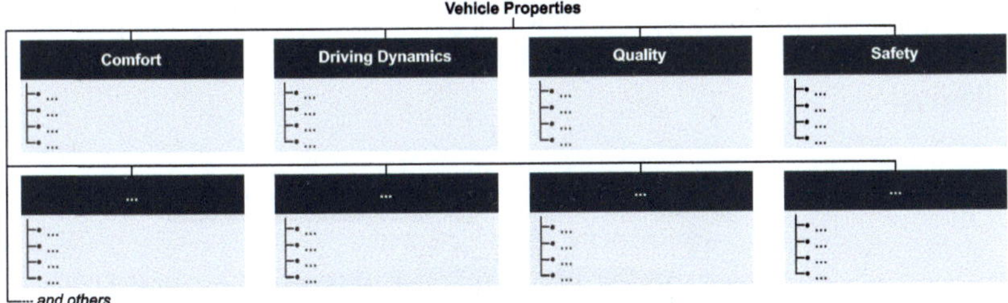

Fig. 3.4 Exemplary overview of the main requirement categories of vehicle properties used in an exemplary automotive OEM

Similar requirement categories can be identified when comparing those three collectives. Nevertheless, due to the degree of detail of an OEM vehicle's property structure, it is challenging to overview vehicle concept overarching requirements. Additionally, the optimized yet traditional automotive development processes need more room for innovations based on novel manufacturing approaches. This condition results, on the one hand, from proven and established conventional manufacturing processes in a mass production context. On the other hand, this relation results from the similarity of predecessor and successor vehicle property structures, as the latter assumes most requirements of the former to minimize development efforts. Therefore, it is necessary to question contemporary vehicle designs regarding the basic requirements of individualized people transportation to project AM potentials on a complete vehicle level. Thus, in the following, a generalized vehicle requirement structure is derived to enable the definition of an idealized vehicle concept.

Fundamental Vehicle Property Structure

The objective of deriving a fundamental vehicle property structure is not to suggest a novel approach to vehicle definition. Moreover, it lays the identification of vehicle concept independent conflicting objectives in the foreground that still represent a challenge for vehicle manufacturers. Consequently, the implementation of AM potentials is most beneficial in areas where, with conventional manufacturing processes, only unsatisfying or sub-optimal compromises are being achieved. The basic requirements defined in this subsection are formulated generally, independent of particular vehicle models, segments, or classes, and from a customer's perspective in the first place.

Based on the three vehicle requirement sets described above, fundamental vehicle properties necessary to cover the most basic demands of individual mobility (personal transport) have been identified. Fig. 3.5 illustrates those six properties abstractly, which are described and listed according to decreasing priority in the following:

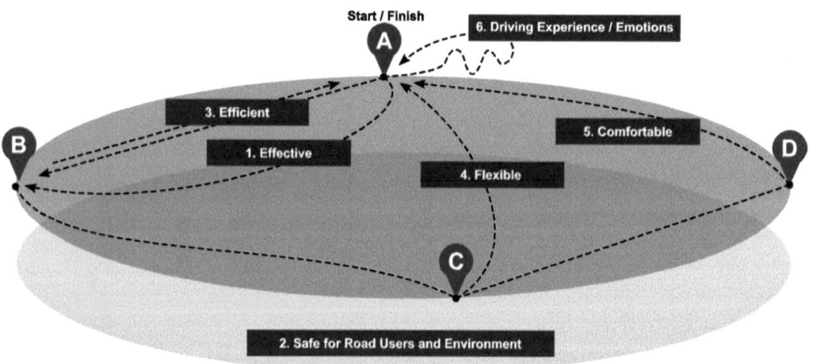

Fig. 3.5 Illustration of fundamental vehicle properties from the perspective of motorized private transport

A vehicle designed for the application area *individual/personal mobility* needs to be ...

1. **Effective:** The desire for individual/personal mobility incorporates the most fundamental objective of arrival at the destination. Therefore, the first most fundamental requirement for automotive vehicles is *vehicle effectiveness*. Thus, all vehicle aspects that contribute to the arrival at the desired destination are subordinated to this category.
2. **Safe:** For an effective arrival at the desired destination, safe circumstances for people and the environment are essential. Therefore, the fundamental requirement for *vehicle safety* consists of all aspects that prevent or inhibit potential harm to active or passive traffic participants and the environment caused by vehicle usage.
3. **Efficient:** The time factor is ordered after safety aspects. Nevertheless, its relevance increases in „accelerated" societies. Hence, vehicle properties that contribute to the minimization of the time and resources necessary for arrival at a desired destination are subordinated to *vehicle efficiency*.
4. **Flexible:** One of the main differentiation points compared to public transport is personal mobility solutions' flexibility. The individual can freely determine and influence the destinations and the route to get there at any time. Thus, the attribute *vehicle flexibility* is a crucial property of automotive vehicles. This category includes all vehicle aspects that enhance its independence, including its suitability to many environments.
5. **Comfortable:** Needs concerning passenger comfort increase according to trip length and duration. This relation can be observed by comparing the comfort features of short and long-distance vehicles. Even though the level of comfort experienced differs between each individual, the importance of *vehicle comfort* for general customer satisfaction increases and thus is considered an additional fundamental vehicle property.
6. **Driving Experience/Emotion:** From its origins, the pleasure of driving has played a central role in automotive history. Despite ongoing mobility changes towards drive train electrification and autonomous driving, it is assumed that pleasurable driving experiences will remain important in upcoming vehicle concepts. Consequently, those vehicle characteristics that favor positive driving experiences are summed up in *driving experience/emotion.*

With this categorization of fundamental vehicle properties in mind, subordinated levels of requirements can be assigned, resulting in a comprehensive vehicle specification. As with many categorizations approaches, single properties might be ambiguous and can be assigned to more than one category. For example, *vehicle's range* can be assigned to vehicle effectiveness (1.), strongly influencing which destinations are within reach. Additionally, the range is also relevant for the vehicle's flexibility (4.) influencing potential areas of operation, e.g., enabling use in areas with low density of charging or refueling infrastructure. Since the fundamental vehicle properties are consecutively prioritized, ambiguous requirements are assigned to the category with higher priority. In this example, vehicle range is assigned to the first fundamental property *vehicle effectiveness*. Following this method and considering the requirements sets analyzed above, the fundamental vehicle requirement structure shown in Fig. 3.6 is derived.

Fig. 3.6 Fundamental property structure of a vehicle concept including general development tendencies. (⬈ maximize, ⬊ minimize, → target value or conflicting objectives)

3.2.2   Conflicts of Objectives in Vehicle Conception

Two objectives are targeted with an idealization of vehicle properties. First, for the following analysis, it is crucial to overview major conflicting objectives in VC. Second, with a comparison between conventional and idealized vehicle properties, optimization potential in VC can be identified. Here, an idealized vehicle is comprehended as a fictitious representation that is not limited by manufacturing restrictions of any kind. Further, conflicting objectives are ignored for the moment in this representation as the optimum of each vehicle property is pursued.

For this purpose, an approach similar to the *House of Quality* as part of the *Quality Function Deployment* (QFD) method is applied (compare [FG13, p. 760; Neh14]). The QFD method often transfers vague yet fundamental customer needs into requirements. Either positive or negative interrelation tendencies between those requirements are also identified. A similar expression of tendencies has been applied to express an idealized vehicle property structure. Contrary to the QFD method, these tendencies focus not on the interrelations between requirements but the interrelation between individual requirements and their optimal fictive representation.

As a result of the application of this principle, the generalized vehicle property structure illustrated in Fig. 3.6 has been complemented with optimization tendencies for each attribute, thereby indicating if the respective attribute should be maximized ($\nearrow$), minimized ($\searrow$) or developed towards a target value ($\rightarrow$) to express an ideal vehicle concept composition. For example, the *operational readiness* (1.1) of any vehicle, including aspects such as reliability, ease of use, and repair, should ideally be maximized ($\nearrow$) to lead to high customer satisfaction during vehicle use.

At this stage, it is challenging to substantiate the property structure by adding further detail. To specify the properties of the structure quantitatively, making it more tangible, a vehicle concept alignment regarding brand values, target group, and scenario of operation is necessary (compare Sect. 2.2.3, Fig. 2.22). Thereby, the general validity of the fundamental property structure for diverse vehicle concepts is lost.

Nevertheless, comparing this idealized vehicle property structure to conventional vehicles facilitates the identification of conflicting objectives in VC. To further identify those conflicting objectives, discrepancies ($\lightning$) between conventional and idealized vehicle property structure are gathered in Table 3.7 based on Fig. 3.6. For example, conventional vehicles require reasonable maintenance and servicing efforts during usage. These efforts represent a discrepancy to optimized *operational readiness* (1.1) vehicle property.

The discrepancies gathered in Table 3.7 represent potential starting points for formulating conflicting objectives in VC. To do so, the definition of conflicting objectives introduced in Sect. 2.2.4 based on EILETZ has been applied (1999). This way, conflict of objectives have been either gathered from the literature ([Eil99; PS16]) or derived from the identified discrepancies from Table 3.7 and summarized as shown in Table 3.8. At least two requirements that impede each other in their fulfillment are opposed in the conflict description. The optimization tendency refers to the first requirement of the conflict and indicates if it should be maximized ($\nearrow$), minimized ($\searrow$), or just be considered ($\rightarrow$). Also, the fundamental requirements illustrated in Fig. 3.5 involved are indicated. As a result of the complexity of a vehicles requirement structure, numerous conflicting objectives can be formulated (compare ESM App. A.2 Table A.2). Those conflicting objectives indicate vehicle aspects that either need to be partially deprioritized or compromises elaborated, leading to customer-relevant vehicle restrictions. Therefore, the approach of this work targets the technical mitigation of such conflicts by applying AM design potentials respectively.

Table 3.7 Extract: Fundamental vehicle property structure, development tendencies ($T,\nearrow,\rightarrow,\searrow$), discrepancies ($\notni$) and resulting categories (**1-4**) between conventional and idealized vehicle representation. (Complete table in ESM App. A.1)

Vehicle Property Structure	T	Discrepancies and Description [$\notni \triangleq$ identified]
1. Effectiveness		
1.1 Operational readiness	$\nearrow$ $\notni$	maintenance and servicing efforts required (**4**)
1.2 Driving (x,y)		
1.2.1 Propulsion	$\nearrow$ $\notni$	Significant origin for trade-offs leading
1.2.2 Maneuverability	$\nearrow$ $\notni$	to substantial vehicle model
1.2.3 Range	$\nearrow$ $\notni$	differentiation at this point (**1**)
1.2.4 Roadworthiness	$\nearrow$ $\notni$	
1.3 Operation		
1.3.1 Transmission of information	$\nearrow$ $\notni$	limited/cumbersome (**3**)
1.3.2 Processing of information	$\nearrow$	
…	… …	…
2. Safety		
2.1 Passenger Safety		
2.1.1 Vehicle control	$\nearrow$ $\notni$	limited field of view (**2,4**)
2.1.2 Provision of safety information	$\nearrow$ $\notni$	limited "health monitoring" (**4**)
2.1.3 Impact protection from kinetic energy	$\nearrow$ $\notni$	Energy dissipation by plastic deformation, driver assistance systems (**4**)
…	… …	…

Table 3.8 Extract: Exemplary conflicts of objectives on complete vehicle level. (Complete table in ESM App. A.2)

T	Conflicting Objectives	Involved Vehicle Property						Source
		Effectiveness	Safety	Efficiency	Flexibility	Comfort	Emotion	
↘	Height of center of gravity ↔ Optimal aerodynamics			•		•		[PS16]
↗	Wind pressure stability ↔ Impact compliance		•					[PS16]
↗	Acoustic comfort ↔ Lightweight design			•		•		[PS16]
↗	Trunk volume ↔ Seating comfort rear seats				•	•		[Eil99]
↗	Rollover protection (convertible) ↔ Seating comfort rear seats		•			•		[Eil99]
→	Diesel variant option ↔ Vibrations			•		•		[Eil99]
↘	Vehicle weight ↔ Comfort features	•		•		•		–
↘	Vehicle weight ↔ Safety features	•	•	•				–
↗	Operational readiness ↔ Maintenance costs and effort	•		•		•		–
↗	Agility ↔ Vehicle model dependency	•		•		•		–
↗	HMI intuitiveness ↔ Complexity of assistance systems	•		•				–
...	...	...	...	...	...	...	...	...

3.2.3   Application Areas of AM Potentials in Vehicle Conception

As a detailed analysis of each identified discrepancy between conventional and idealized vehicle properties and the derived conflicting objectives in the previous subsection would exceed the scope of this work, these aspects have been matched to four key application areas of AM in VC. The application of AM design potentials to these areas facilitates mitigation of conflicts in VC consequently and thus results in tangible customer benefits. This categorization is based on a comparison between identified discrepancies (Table 3.7) and the holistic AM application possibilities implied in the semantic network of AM design potentials by KUMKE ET AL. (2018) introduced in the fundamental Sect. 2.3.3 Fig. 2.31. These four application areas are numbered in Table 3.7 (**1–4**), illustrated in Fig. 3.7 and will be elaborated on in further detail.

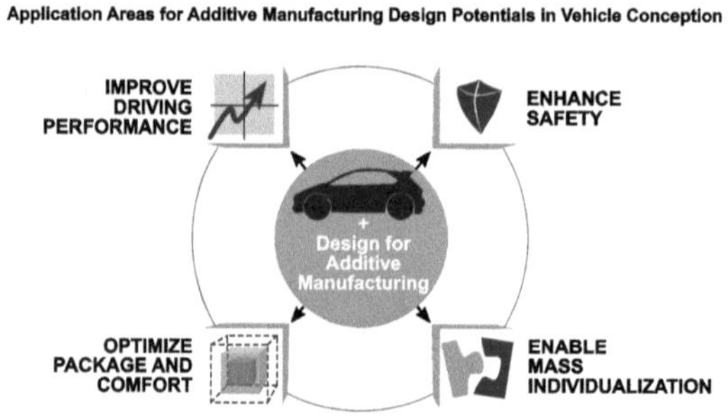

Fig. 3.7 Key application areas of AM potentials for vehicle conception

Due to the interdependencies between the application areas in Fig. 3.7, the aim of AM applications in these areas is not the optimization of the respective area individually. This objective would also be technically feasible with conventional manufacturing technologies by limiting or deprioritizing one or more of the remaining areas. Moreover, optimizing one respective area with DfAM is targeted without diminishing the remaining ones. This way, this approach stays focused on added value enabled by AM, challenging solutions not feasible with conventional manufacturing methods. This principle applies to the following descriptions:

1. **Improve driving performance:** The first application area focuses on minimizing driving resistances and maximizing energy conversion efficiencies in a vehicle. AM applications that contribute to this purpose belong to this area. Here, key vehicle aspects related to driving resistances are analyzed to minimize the major components, e.g., sprung, unsprung, and rotational vehicle masses (compare Eq.2.1 to 2.3 in Sect. 2.2.1).

2. **Optimize package and comfort:** Vehicle sections with unused or wasted design space are targeted in this application area. Here, design space analyses are the starting point with a particular interest in continuous vehicle dimensions (Fig. 2.24) and sections of conventional vehicles subject to, e.g., body-cavity-sealing. Since optimizing space utilization in a vehicle is often strongly linked to the ergonomic comfort of occupants, comfort-relevant AM application ideas are also assigned to this category.

3. **Enable mass individualization:** This identified application area is related to the diversity of customer needs. Today only limited individualization options can be offered to the customer due to the dependence of vehicle properties on conflicting objectives in their requirements and the dependence on flexibility-restricted

manufacturing processes in mass production. With the implementation of AM processes, additional individualization options can be realized, enabling not yet feasible customer benefits enhancing, e.g., vehicle's flexibility and emotional properties. As vehicle styling is considered subjective in customers' perception, styling-related AM applications are also assigned to this application area.

4. **Enhance safety:** The conflicting objective of vehicle safety and efficiency focuses on the fourth potential application area for AM. In conventional vehicles, additional passenger safety measures lead to increased vehicle weight diminishing driving performance, and in some cases, decreased impact safety for collision partners. For example, optimizing and redesigning load-bearing structures might enhance vehicle safety properties without additional material counteracting the „weight spiral".

With the fundamental vehicle properties defined, the conflicting objectives in VC gathered, and the main application areas for AM in VC identified, conceptual ideas are generated for each application area in the next step. First, conflicting objectives of the respective application area are analyzed to determine their key factors. A conceptual phase follows in which potential solutions are generated employing DfAM methods introduced in Sect. 2.3, e.g., in Table 2.12. This step is analog to step three of the DfAM framework overview suggested by KUMKE ET AL. (2016) [KWV16, p. 12]. Also, AM-unspecific approaches such as creativity methods (Table 2.3), topology optimization (for more context on *topology optimization* please compare [Chr+15; Fra18; DPL21; Li+21; Zhu+21; Bar22]) , bionic databases [Emm+11], and ideation workshops with experts can be helpful tools during the development of novel ideas for optimization of conflicting objectives in VC. Afterwards, sub-concepts visualizing the generated ideas have been developed. Following this procedure, and by applying approaches and methods, such as topology analysis, semantic network of AM design potentials, bionic data bases, mind mapping, creative workshops, and consulting experts, the application principles and sub-concepts gathered in Table 3.9 have been generated (compare ESM App. A.3). Here, the definition of AM potentials as a combination of AM design principle and value proposition (Sect. 2.3.3) has been applied to clarify relevant AM aspects for each context.

The sub-concepts gathered in Table 3.9 should be considered exemplary. Most application principles are also applicable to other vehicle sections or components. Further, the examples generated only visualize individual technical application principles yet do not represent the impact of AM design potentials on a complete vehicle concept. This aspect is elaborated further in the following chapters.

Specific technical concerns regarding the feasibility of each application principle are also listed in Table 3.9. However, at this stage, a viable assessment of the economic efficiency of each approach is not expedient as feasibility, target segment, AM process, and other aspects are not yet defined. Therefore, cost-effectiveness remains a concern overarching all application principles of Table 3.9.

The ideas gathered in Table 3.9 indicate that unconventional approaches to vehicle design arise when methodically considering AM design potentials and conflicting objectives in vehicle conception. When considering combining those principles, interrelations between AM potentials become relevant, particularly when multiple AM value propositions are targeted in the same product. As mentioned in Sect. 2.4.3, contemporary work in DfAM does not address AM potential interdependencies. For this reason, this aspect is further clarified in the following section.

Table 3.9 Extract: Overview of generated AM application principles in vehicle conception considering the four major areas identified in Fig. 3.7. (Complete version in ESM App. A.4 Table A.3)

1. Improve driving performance

1.1...

1.2 Load bearing vehicle components (structurally integrated) by functional complexity

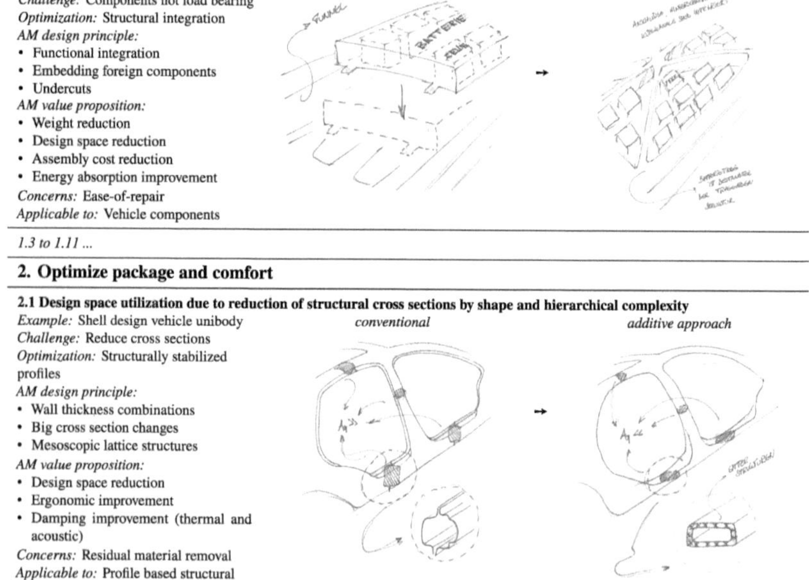

Example: Vehicle battery integration
Challenge: Components not load bearing
Optimization: Structural integration
AM design principle:
- Functional integration
- Embedding foreign components
- Undercuts
AM value proposition:
- Weight reduction
- Design space reduction
- Assembly cost reduction
- Energy absorption improvement
Concerns: Ease-of-repair
Applicable to: Vehicle components

1.3 to 1.11 ...

2. Optimize package and comfort

2.1 Design space utilization due to reduction of structural cross sections by shape and hierarchical complexity

Example: Shell design vehicle unibody
Challenge: Reduce cross sections
Optimization: Structurally stabilized profiles
AM design principle:
- Wall thickness combinations
- Big cross section changes
- Mesoscopic lattice structures
AM value proposition:
- Design space reduction
- Ergonomic improvement
- Damping improvement (thermal and acoustic)
Concerns: Residual material removal
Applicable to: Profile based structural components

(continued)

Table 3.9 (continued)

2.2 to 2.5 …		

3. Enable mass individualization

3.1 Individualized vehicle dimensions by toolless manufacturing

Example: Derivative body structure
Challenge: Enable vehicle derivatives
Optimization: Tool independent space frame design
AM design principle:
- Big cross section changes
- No draft angles
- Part consolidation

AM value proposition:
- Variant costs reduction
- Logistics cost reduction
- Reduced number of joints

Concerns: Crash worthiness
Applicable to: Vehicle sections or components

3.2 …

4. Enhance safety

4.1 Geometrical ductility of structures by shape and functional complexity

Example: Longitudinal member front
Challenge: Elastic energy absorption
Optimization: Compliant structures
AM design principle:
- Freeform geometry
- Big cross section changes
- Macroscopic lattice structures

AM value proposition:
- Energy application improvement
- Ease of assembly
- Repair cost reduction

Concerns: Deformation limited by package
Applicable to: Components with directional ductility requirements

4.2 to 4.3 …

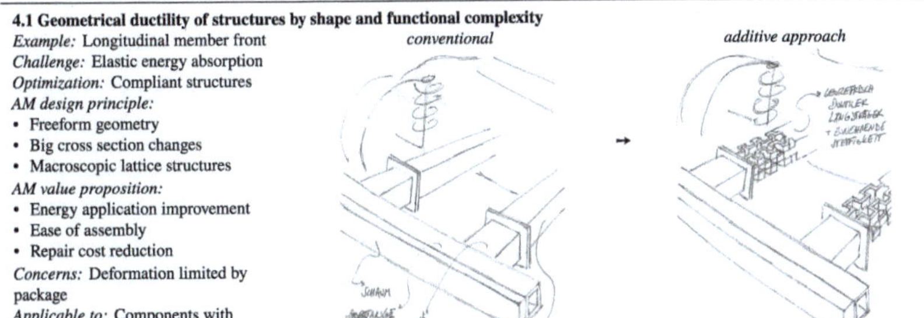

3.3   Interrelations and Conflicts Between AM Design Potentials

Analog to the field of research DfAM, this section targets to comprehend facets of AM design potentials as well as provide knowledge about relevant restrictions in this context. Here, a systematical pair-by-pair comparison [Neh14] is applied to identify interdependencies between AM-design potentials as they are defined in the semantic network of AM design potentials (Fig. 2.31). A purposeful application approach of AM leads to one or multiple optimized product properties, also known as value propositions (Sect. 2.3.3). In the automotive context, this might be, e.g., reduced vehicle weight, improved heat transfer, improved ergonomics, and others (compare in Table 3.9). In relation to the complexity of a product, engineers strive to optimize numerous product properties simultaneously. By using the semantic network of AM potentials [Kum+18], it is possible to identify which design principles are directly linked to the targeted value propositions and could be implemented into the design. In this approach, though, it is not considered that the implementation of multiple design principles might negatively affect certain value propositions. For example, if mesoscopic lattice structures are used to reduce material usage and, therefore, weight in a constant design space, the stiffness of the topology will decrease. Otherwise, to achieve the same stiffness as without lattice structures, a larger design space is needed to compensate for the lower material usage. Therefore, a negative interrelation between the value proposition *weight saving* and *design space reduction* might occur with the use of lattice structures. Engineers should be aware of such relations in order to foresee the negative side effects of certain AM design principles. This way, it is possible for design engineers to prioritize and decide if the resulting product properties still meet the product requirements [Fuc+20].

To clarify and address these potentially negative interrelations, a *matrix of conflicting AM potentials* has been developed. During the identification of dependent AM potentials for this matrix, it becomes clear that AM design principles themselves (e.g., internal walls/ribs, honeycombs, surface textures, ...) do not necessarily contradict each other just based on physical principles. Yet, between the AM value propositions (e.g., weight saving, design space reduction, improved heat transfer, ...) resulting from AM design principles, physical mutual dependencies can be identified (Fig. 3.8).

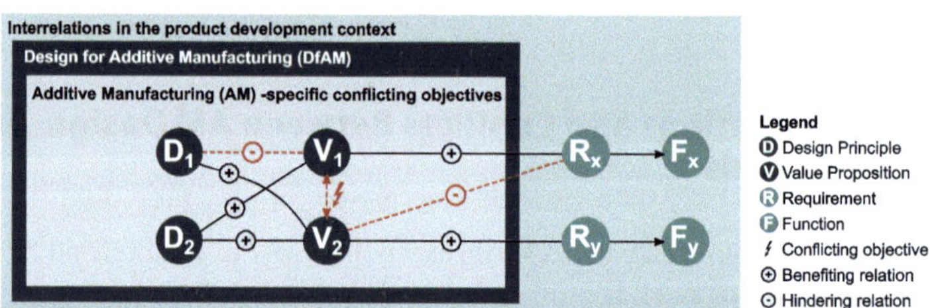

Fig. 3.8 Schematic display of positive and negative interrelations between AM value propositions, their dependencies with AM design principles, and general product requirements [Fuc+20]

For a holistic analysis of conflicting AM potentials, all product-related value propositions of the semantic network of AM potentials [Kum+18] are opposed to each other in a pair-by-pair comparison as illustrated in Fig. 3.9. In the next step, the physical compatibility of each pair of AM value propositions matrix is questioned.

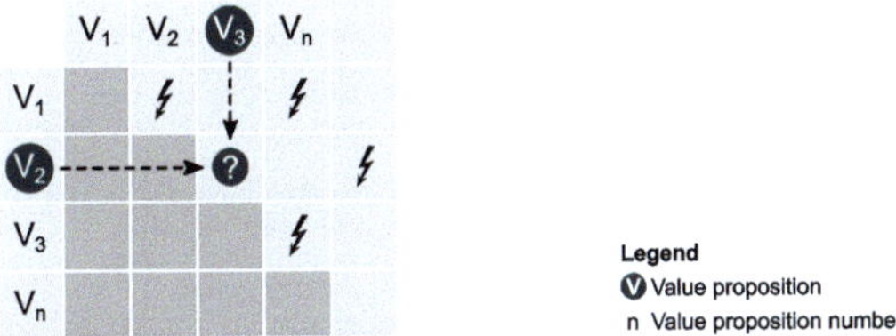

Fig. 3.9 Schematic layout of the pair-by-pair comparison of value propositions applied in the matrix of conflicting AM potentials [Fuc+20]

If a conflict between two AM value propositions is identified, it is described in the respective overlapping cell of the matrix. Further, the AM design principles related to each conflicting value proposition are identified. With the aim not just to point out conflicts but also to help with their mitigation, lastly alternative AM design principles are suggested. Table 3.10 shows an extract of the matrix of conflicting AM potentials assessing the value proposition *design space reduction* in more detail.

However, the content of the suggested matrix of conflicting AM potentials is not exhaustive and not yet quantified. Further, with the aspiration to consider AM potentials holistically, the semantic network of AM potentials by KUMKE ET AL. (2018), as well as the developed matrix of conflicting AM potentials, do not consider process-specific restrictions. As the AM-related design freedom differs strongly between processes, the identified interrelations between conflicting AM potentials are not equally relevant for each AM process.

Concluding, the matrix of conflicting AM potentials enables design engineers to obtain additional information in regard to AM value propositions of interest and, in particular, regarding their combination. Thereby, conflicting objectives are pointed out, which indicate possible problems and sensitize the engineer accordingly. Additionally, in the matrix, alternative design potentials are suggested, assisting users without profound DfAM knowledge during conflict mitigation. Exemplary use cases of the matrix are explained in more detail in FUCHS ET AL. (2020) [Fuc+20].

Table 3.10 Extract: Matrix of conflicting AM potentials based on [Fuc+20]. (V = value proposition; D = design principle; $\frac{1}{2}$ = conflict identified; complete table in ESM App. A.5)

	V_1:	V_2: Weight reduction	V_3: Improve energy conversion/transport	V_4: Improve energy absorption	...	V_{12}: Improve insulation
conflict		$\frac{1}{2}$	$\frac{1}{2}$	$\frac{1}{2}$	...	$\frac{1}{2}$
cause		Mesoscopic structures lead to weight reduction yet increased design space at constant stiffness	Design space reduction can lead to flow losses due to reduced curvature radii	Design space reduction can lead to decreased energy absorption due to decreased contact area	...	Mesoscopic structures can improve insulation yet increase design space
D related to V_1		Lattice structures, honey combs, irregular porous structures	Freeform geometries in cavities and channels, reduced number of joints, part consolidation	Lattice structures	...	Lattice structures, honey combs, irregular porous structures
D related to conflict partner V_n				Part consolidation, small gaps between features, big cross-section changes	...	
Alternative D		Topology optimization, lattice structures, wall thickness combinations, different materials in one part		Material complexity	...	Different materials in one part, wall thickness combinations, surface textures
V_2			...	...	...	...
V_3				...	...	...

(Row label for the whole block on the left: *V_1: Design space reduction*)

3.4   Summary and Conclusion

To synthesize the potential benefits of AM on VC, the state-of-the-art of AM applications on the complete vehicle level have been analyzed and structured in this chapter. Here, five different approaches to additively manufactured vehicles are distinguishable: *unique show cars (one-offs), individualizable vehicle concepts, low requirement vehicles, high-performance vehicles*, and *generative concepts*. Additional information regarding AM value propositions in focus, such as lightweight design, toolless manufacturing, assembly cost reduction, and others, as well as the addressed technical sections (vehicle penetration level) and the AM processes used, has also been provided. Moreover, an overview of the conventional vehicle body design approaches already represented in the state-of-the-art of additively manufactured vehicle bodies has been provided.

Based on the technical complexity of automotive vehicles and the diversity of potentials assigned to AM, systematic DfAM approaches to this context are considered necessary, yet not found in state-of-the-art applications. Therefore, fundamental vehicle properties and their conflicting objectives in VC have been analyzed, and four main application areas for AM potentials in VC identified. Based on those main application areas along with DfAM methods, twenty-two technical application principles have been developed. With exemplary sub-concepts, those application principles are visualized, illustrating potential vehicle optimizations powered by AM design freedom. Thereby, those sub-concepts take advantage of different AM design principles to achieve multiple value propositions with customer relevance.

Further, as not yet addressed in contemporary work in the field of DfAM, possible interrelations between AM potentials are examined in the last subsection of this chapter. Here, a pair-by-pair comparison of all product-related AM value-propositions of the semantic network of AM potentials has been developed, resulting in a matrix of physically conflicting AM potentials. This matrix represents a methodological tool that provides additional information for engineers during the early design phase, indicating possible collateral effects when striving for multiple AM value propositions in one product.

As summarized in the state-of-the-art, there are no holistic approaches to implementing AM potentials on the complete vehicle level. Each of the AM vehicle concepts introduced in this chapter focuses on a single or a few of the multitude of value propositions assigned to AM. Here, a need for a systematic approach is identified to leverage AM potentials holistically. Thus, an idealized vehicle property structure has been derived, enabling the identification of optimization potential considering all fundamental vehicle aspects. By applying AM potentials to those vehicle optimizations, added customer value can be achieved, not yet accessible with conventional manufacturing methods. Even though numerous innovation ideas can be generated addressing the four identified main application areas of AM potentials in VC, additional research questions arise on how to deal with increased development complexity. A random combination of differing AM value propositions and their related AM design potentials turns out not to be trivial. Their mutual dependencies need to be taken into account, possibly restricting holistic DfAM approaches.

With these findings in mind, it becomes clear that a detailed partial analysis of the AM application principles is necessary to assess their feasibility and estimate their possible impact on VC. Those partial analysis are the focus of the following chapter.

Elaboration of Specific Automotive AM-Potential Applications

4

In the previous chapter, four general application areas for additive manufacturing (AM) potentials have been derived based on conflicting objectives in vehicle conception (VC). Further, principle conceptual innovation ideas based on AM value propositions and design for additive manufacturing (DfAM) methods addressing the four application areas have been generated. Mutual dependencies between AM value propositions have been methodically elaborated. Nevertheless, major technical development challenges regarding these application principles' feasibility arise before their detailed development. Thus, this chapter elaborates on these fundamental challenges to assess the feasibility of diverse AM application principles in the automotive context.

This chapter shifts the focus of this thesis from the complete vehicle level down to specific AM applications. This change in focus enables tangible studies on different challenges in this context (compare Sect. 1.4). These challenges address different aspects of increased development complexity occurring when optimizing designs to leverage AM design and manufacturing freedom. Here, two major development challenges relevant to each AM application area in VC (compare Sect. 3.2.3) are covered. First, based on the toolless characteristic of AM processes, Sect. 4.1 addresses variant complexity elaborating on how to develop mass-individualized vehicle components. Second, in Sect. 4.2, challenges regarding the development of functionally complex components are analyzed, that result when taking advantage of the interdisciplinary characteristic of DfAM-related functional integration potential.

Supplementary Information The online version contains supplementary material available at https://doi.org/10.1007/978-3-658-49677-7_4.

## 4.1	Engineering Design and Functional Assurance of Mass-Individualized Vehicle Components

As introduced in the fundamentals Sect. 2.1.3, *Mass Individualization* (MI) can be considered an overall market trend [Kor+15] yet it induces additional complexity to design methodology [LRZ06]. Simultaneously reducing product variants of automotive OEMs is one key factor in reducing corporate complexity and costs. As KABASAKAL ET AL. (2017) emphasize, the strategic advantage of customization is greater where product differentiation is hard but crucial, mentioning a reason why the automotive industry still equips car models with numerous customization options [Kab+17]. However, state-of-the-art customization options feasible for OEMs are often restricted to changes in trim colors, materials, and the final combination of chosen vehicle specifications, such as drive train options and additional functions. A customer-specific variation of vehicle shapes and part geometry is only known to small series premium manufacturers due to the strong dependence of automotive mass production on tool-based manufacturing processes. The tools used, e.g., for sheet metal deep drawing, are costly and, at least for the volume segment, only amortize in mass production of identical parts.

As toolless manufacturing technologies diminish the conflict of interest between mass production and crucial product differentiation, due to their nearly lot-size-independent characteristic, AM processes are strongly assigned to MI in the literature [Tuc+08; Ber12; Con+14; Zhe+17; Kor21]. Although no additional tools or tool changes are necessary for AM to produce different variations of a part's shape, other aspects of variant-related corporate complexity remain challenging. With no suitable approaches to these development challenges, implementing AM-related mass individualized applications in the automotive series production will remain to defy. Thus, this subsection focuses on challenges regarding AM-enabled mass product individualization in the automotive context, focusing on shape individualization.

As defined in Sect. 2.1.3, an individualized product is a customer-specified product based on unique customer attributes, such as anthropometry, individual preferences, experiences, memories, and others. In the automotive industry, at least four different product individualization approaches can be distinguished (Fig. 4.1). These four categories describe the type of additional value for the customer intending to enhance the customer-vehicle-identification and have been identified by grouping similar individualization applications of the same kind.

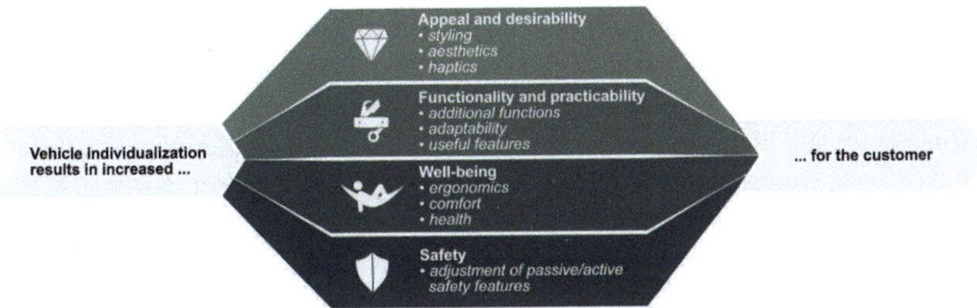

Fig. 4.1 Overview of identified approaches to AM-related vehicle individualization focused on additional customer value

These four categories represent possible approaches to vehicle individualization and show different motivations (Table 4.1). Supposed that the type of added value correlates to the preferences of the targeted customer group, increased customer-

Table 4.1 Identified approaches to AM-related vehicle individualization explained

Description	*Example*
1. Appeal, desirability, and differentiation: Individualization options are often targeted to increase customer product appeal and desirability. This category focuses on the vehicle's exterior and interior styling, aesthetics, and haptics. One example is individualized patterns and structures on trim elements (*see door trim pattern based on city topology from bird's-eye perspective on the right*).	
2. Functionality and practicability: The actual functional aspects of a vehicle might also be adapted to the customer's needs. Either regarding, e.g., individualized storage options, mechanical interfaces, or individually positioned or designed operating elements of the vehicle (*compare modular center console containing individualized storage composition on the right* [Fuc+20]).	
3. Well-being: Individualized vehicle comfort, ergonomics, and health-related aspects fit this third category. Here, a bespoke anthropometry-based seat shape and or seating position, e.g., leads to additional customer benefits not yet feasible with off-the-shelf components (*see ergonomic interior door handle shaped according to customer's unique anthropometry on the right*).	
4. Safety: Lastly, also customer-individual safety-relevant vehicle alterations are imaginable with toolless manufacturing, such as anthropometry-specific vehicle body reinforcements, individual pillar positioning for increased visibility, or bespoke seats for improved crash protection and so on (*compare individualized seat inserts based on customer's unique anthropometry* [Por20]).	

vehicle identification can be achieved and, therefore, a differentiation from similar yet non-individualized competitor products. Overarching those individualization categories, customer involvement in a vehicle's composition or design leads to a *self-made* feeling that associates customers with their vehicle. Additionally, differing customer groups, such as private owners, corporate and fleet customers, and automotive dealers, are relevant for vehicle individualization.

To enable individualized vehicle aspects with AM, the following development challenges have been identified:

- Product specification and order process (information input)
- Multi-variant engineering design
- Multi-variant functional and quality assurance
- Brand identity (design language)

If not solved or diminished, these challenges propel corporate complexity and inhibit positive automotive business cases in the volume segment. As variant-related product complexity is well-described in the literature (Sect. 2.1.3), the following challenges focus on particularly challenging aspects when individualizing a part's shape with AM-related design freedom in mind.

The points mentioned above emphasize that even with reduced variant-related manufacturing complexity due to toolless AM technologies, the remaining variant-related aspects of product development need to be coped with to leverage this AM potential. The applications introduced below further clarify these challenges and the product individualization categories illustrated in Fig. 4.1.

As shown in Fig. 2.14 an individual customer attribute is a key element for product specification differentiating MI from MC (compare also definitions 10 and 11). At the beginning of the specification process, one major question arises: „On what kind of individual customer attribute should the mass individualized automotive product be based?" This question leads to various possible answers as individual customers are manifold. Therefore, Table 4.2 summarizes a few customer attributes that might be of particular relevance for a certain targeted group of customers and can contribute to customer-vehicle identification.

Table 4.2 shows that at least two major groups can be distinguished when gathering individual customer attributes. The first one is the uniqueness of the individual in all its facets. Here, unique aspects such as, e.g., anthropometry of the body, conviction, taste, memories, and needs are grouped. The second one might be the identification/differentiation of a certain individual with/from his social environment. This group consists of attributes expressing affiliation or non-affiliation to

Table 4.2 Exemplary individual customer attributes as point of reference for mass individualized products

Uniqueness of the individual	Identification/ differentiation of the environment	Demonstration possibilities
• Physical characteristics (fingerprint, DNA, anthropometry, voice, iris) • Skill set, aspirations, values • Opinion, prejudices, knowledge, culture, origin, religion • Preferences (colors, shapes, sound, music, need for tidiness) • Memories, experiences • Desires, fears (need for privacy, protection) • Perception (warmth, coldness, motion sickness) • Environment, relationships, pets • Image • Day-to-day life, biorhythm • Motivation, drive • Geographical references	• Vehicle model, brand, status • Likings • Memories • Being part of a group • Colors and shapes • Settings • Subjective ergonomics and well-being • Relation to family and friends	• Add-on components • Display frames • Car wrap • Ambient light, light projections • Accessories (key fob) • Language settings • Tattoos, piercings • Surfaces and textures • Augmented reality, position-related information • Generative design according to individual specifications • Vehicle appearance similar to own style • Implementation of foreign materials, textures, and shapes as cross-reference

a certain social group, e.g., with a certain hobby, homeland, cultural, or political statement.

With the relevant attributes of the targeted customer group of choice identified, the second question arises: „How could those customer attributes be related to the product's shape?". Considering the shape is the major characteristic of a product that can be easily altered with AM processes, additional characteristics such as color, finishing, and others are not in the focus of this section, yet not less relevant apart from the AM context. The following Fig. 4.2 categorizes six different alteration levels of a part's shape.

Beginning with *smooth surfaces* Fig. 4.2 illustrates that a part's surface and shape can be altered in increasing complexity levels up to its entire *topology* to incorporate individual customer attributes. Smooth surfaces are suitable for overprinting or treatment with many finishing processes. AM gains relevance from the second surface complexity level when implementing *engraved or embossed* details, such as individual wording and image motifs. Further, surface *graining and texturing* can additionally be individualized according to customers' preferences, e.g., with the emblem of their favorite soccer club. A *3D texture* describes surfaces deeply textured beyond the mesoscopic level leading to the next complexity level *3D structures*

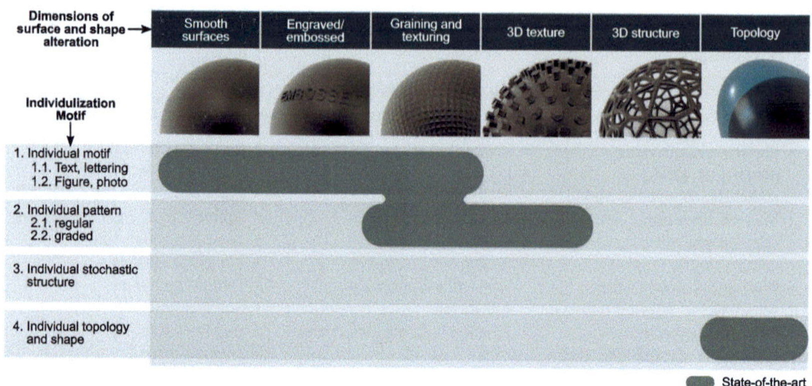

Fig. 4.2 Different levels of shape alteration relevant in the AM and MI context

that also consist of undercutting self-supporting elements such as beams and struts. Latter surface complexity level might also be extended to *4D structures* referring to surfaces with integrated flexible, adaptable, or moving elements that can be actuated. Lastly, beyond the part's surface, its entire *topology* can also be individually altered when additively manufactured, e.g., an individual ergonomic seat. Figure 4.2 indicates that state-of-the-art approaches do not yet cover all possible combinations of shape alteration approaches and individualization motifs.

Considering an individual attribute relevant to the customer and the levels of shape alteration possible with DfAM, the fundamentals for individualizing vehicle components are set. To demonstrate this approach, an individualized door trim application is introduced next.

4.1.1 Mass-Individualized Vehicle Appeal and Desirability

The findings described in the section above are applied to an automotive part to elaborate on the feasibility of mass-individualized products. Here, the first individualization approach of Fig. 4.1 *appeal and desirability* is targeted. This use case aims to familiarize the vehicle's interior for the customer in an individual and aesthetic manner. For this purpose, the exemplary individual customer attribute *hometown affiliation* is chosen. Additionally, to generate an intriguing yet subtle impression, this attribute shall be implemented into a 3D texture (Fig. 4.2). A door trim insert was chosen as a demonstrator, representing other decorative interior parts.

A creative and technical design challenge arises when implementing a customer's hometown affiliation into a 3D texture on a door trim insert. One possible approach is utilizing the bird's-eye view of a particularly important section of the customer's hometown as a 3D texture for the insert. This way, an intriguing yet individual appearance might result. Besides designing the single part itself, the major challenge here is elaborating a semi-automated workflow that enables creating those individual inserts in a MI context.

A key technique in this context is parametric design. A parametric associative geometry representation in CAD enables the change of input parameters within mathematical boundaries, automatically updating the part's geometry accordingly with no need for a manual redesign [Pot+05, p. 751; SD12, p. 38]. The limits to conventional parametric design will be explained in more detail in Sect. 4.1.3. However, an uprising design principle for geometry manipulation based on parametric associative design complements the three conventional approaches explained in Sect. 2.1.2, is *Visual Scripting* (Fig. 2.10).

Definition 25: *Visual Scripting* – In the context of engineering design *visual scripting* represents an alternative approach to geometry generation, manipulation or simulation in CAD/CAE. Here, modules representing a certain operation or function get interlinked with input and output connections resulting in a logical blueprint for a parametric associative geometry. According to its name, visual scripting is a logical programming approach without coding or manual modeling in the first place, thereby reducing design time for complex geometries significantly [Car+21, p. 6] (compare the list of visual scripting tools in [Mat23, p. 3]).

Applying this approach, the design challenge of an individualized decorative door trim insert was solved by combining the functionalities of *Open Street Map* with *Grasshopper* (a *Rhinoceros* extension for visual scripting) and *Catia V5* according to the following steps:

1 Open Street Map
   1.1 Export outlines of buildings in the town section of interest
2 Rhinoceros 7 and Grasshopper
   2.1 Import into the visual script as major customer-specified input
   2.2 Shape and outline of the part are non-variable inputs to the script

   2.3 Level out outline contours of part to 2D

   2.4 The building outlines of the town section of choice are positioned in the level outlines of the part as chosen by the customer
The following steps are carried out automatically by the visual script

   2.5 Split building outlines inside of part contour from outside lines

   2.6 Close open outlines by the splitting operation

   2.7 Project level contours back onto the initially curved base surface of the part

   2.8 Extrude outlines perpendicular to the base surface up to at least part depth

3 Catia V5

   3.1 Split base surface of the part with extruded building surfaces

   3.2 Split extruded building surfaces with the top surface of the part

   3.3 Join surface segments to one water-tight B-Rep

   3.4 Tessellate and export STL-file for AM

4 Slicer of choice

   4.1 Slice geometry file and prepare print job

The logical blueprint resulting from this step-by-step process is shown in Fig. 4.3. Once developed, the input graphics of the customer's hometown affiliation can be altered, and the door trim geometry will update itself accordingly. However, the limits of the geometry kernel and the mathematical representation also apply to this design approach. Nevertheless, when ensuring the quality of input data, e.g., „water tight" building outlines a robust parametric design blueprint is possible reducing design effort for individualization applications significantly.

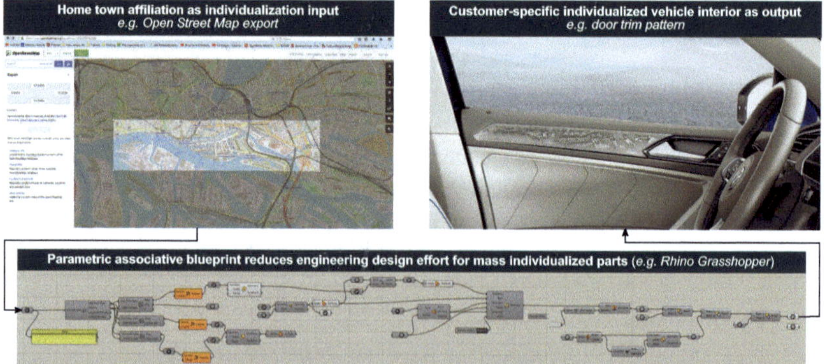

Fig. 4.3 Visual script overview of a blueprint for an individualized automotive door trim design with hometown affiliation

The resulting door trim insert shown in Fig. 4.3, represents a subtle yet individual decoration of the vehicle's interior. Even though the resulting 3D texture might be complex to manufacture, depending on the chosen town section and scale, AM is not crucial for the manufacturability of this single part. This example might also be manufactured using conventional tool-based manufacturing processes like injection molding. AM is essential for the toolless manufacturability of the numerous variants of this application generated with the parametric design approach introduced above.

The finished prototype was manufactured with the selective laser sintering (SLS) process (Fig. 4.4). For smooth surfaces, alternative processes such as SLA and DLP are suggested (compare AM-process-overview in Fig. 2.30). Thereby, UV-light-resilience considering UV stability needs to be enhanced with adequate finishing steps such as, e.g. spray painting. As this application only covers an aesthetic approach to vehicle individualization, a second use case focusing on individualized functionality is introduced in the following.

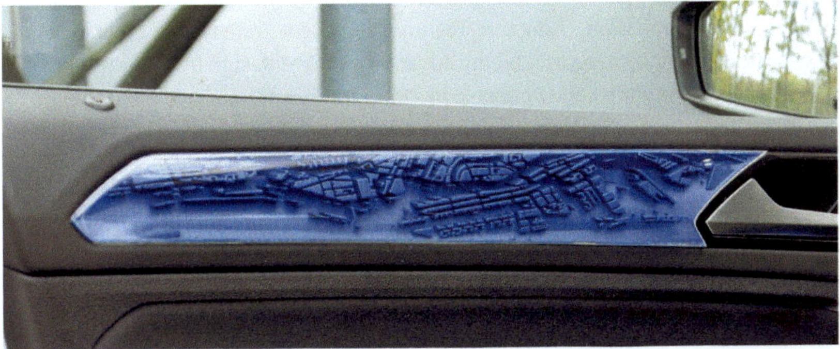

Fig. 4.4 Finished customer individualized automotive door trim insert implementing home town affiliation as decorative 3D surface structure

4.1.2   Mass-Individualized Vehicle Functionality and Practicability

Functionality and practicability are a second approach to vehicle individualization as customer preferences differ strongly in this regard (Fig. 4.1). WAGNER ET AL. (2016) assess different approaches to functional interior customization addressing customer-individual storage behavior and suggests to implement such options in automotive interior layout [Wag+16, p. 884]. Even though rapid technologies are mentioned as enablers for such products and the test samples were additively manufactured with SLS process in WAGNER ET AL.'s work, details on additional possibilities to functional individualization due to AM design freedom are not further elaborated.

As mentioned in FUCHS ET AL. (2020), AM enables additional functional individualization options that exceed the combination of preconfigured yet standardized modules. AM introduces shape variation to functional individualization, increasing the solution space significantly. Also, in this context, implementing individual customer attributes differentiates functional customization and individualization. The following use case of an individualized center console in the vehicle's interior was identified as a suitable example to emphasize those aspects. The design objective is to implement functional individualization options for the customer in a center console, creating specific storage for personal belongings and enhancing customer-vehicle identification [Fuc+20].

The options developed for the center console concept can be distinguished into passive and adaptive features. Passive features are individualized during the specification process and additively manufactured due to the uniqueness of their shape and feature combination. Adaptive features represent individually specified yet adjustable elements during use, taking advantage of AM design freedom. Figure 4.5 illustrates an exemplary center console incorporating individualized passive storage partitions and adaptive holders for cups, bottles, smartphones, and other belongings

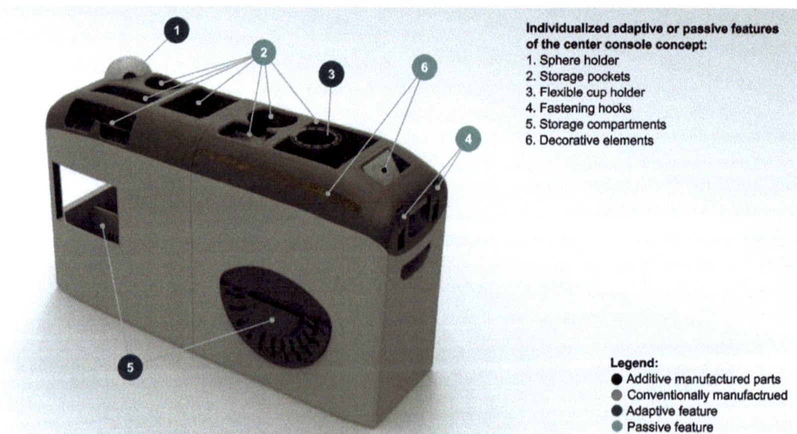

Fig. 4.5 Concept for a customer-individualized automotive center console incorporating individual passive and adaptive features enabled by AM based on [Fuc+20, p. 237]

[Fuc+20]. The fundamental individualization approach in this concept is based on implementing the possibilities of parametric associative design into advanced specifications tools. This way, the customer is enabled to individualize the shape and position of the storage partitions and the additional desired features. The individual customer attributes in focus here are the customer-specific preferences regarding the belongings to be stored and the „do it yourself" aspect of the specification process [Fuc+20].

The adaptive features in this use case are realized by applying AM design principles from functional and hierarchical complexity. For instance, the adaptive smartphone holder shown in more detail in Fig. 4.6 is additively manufactured interlaced in one build job, although its components allow for relative movements. It is based on a spherical geometry containing several cutouts specific to the customer's belongings. The sphere can be rotated to position the cutouts as desired. A spring element retains the sphere in place when preloaded [Fuc+20].

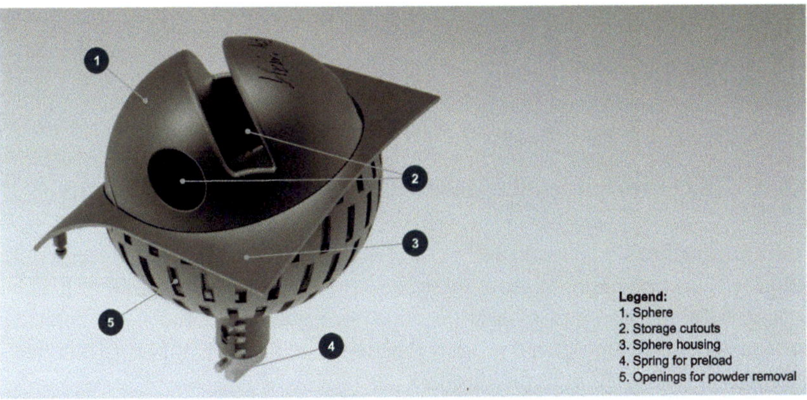

Fig. 4.6 Exemplary adaptive individualization feature of a customer-specific automotive center console concept [Fuc+20, p. 237; KF21]

However, with the applications introduced in the context of aesthetic and functional vehicle individualization, it becomes clear that customer-specific and unique geometries are not only an engineering design yet a development challenge too. This relation arises from interdependencies between customer-specified and general product requirements. Those interdependencies become clear when individualization options compromise product functionalities and thereby the fulfillment of product requirements [LRZ06]. WAGNER ET AL. (2016), e.g., mention that automotive engineers might have safety concerns regarding additional customer-specific functional features in the vehicle's interior due to additional interfering contours in the case of a crash.

At this point, it is important to differentiate the terms *functional individualization* and *function impairing individualization.*

> **Definition 26:** *Functional Individualization* – The customer-specified selection, design or combination of product features or characteristics, which lead to increased, e.g., product practicality for the customer [Fuc+20, p. 236].

> **Definition 27:** *Function Impairing Individualization* – The customer-specified selection, design or combination of product features or characteristics, which compromise hierarchically superordinate product functions. In this case, a certain individualization aspect conflicts with a higher-level product requirement. [Fuc+20, p. 236].

4.1.3 Mass-Individualized Well-Being of Vehicle Occupants

The individualization approach *well-being*, focused in this subsection, includes individual aspects that might improve the customer's ergonomics, comfort, or health. Additionally, the question of how to cope with mutual dependencies between customer-specified and general product requirements (function impairing individualization) is elaborated in further detail.

For this purpose, a use case was chosen that fulfills a particular function yet is tangible enough for this engineering design challenge. The application should also represent comfort-relevant vehicle components and be assignable to a certain individual customer attribute. Thus, the interior door handle was chosen as an exemplary automotive part. Moreover, the approach elaborated applies to additional comfort-relevant parts such as seats, steering wheels, and armrests.

As approaches for functional assurance of individualized AM applications have yet to be further addressed in the literature, this analysis aims to develop an approach for a customer-specified individual interior door handle while assuring its functional scope. In addition, this approach should contribute to reducing variant-related development complexity.

Table 4.3 illustrates and summarizes the functional scope of the conventional embodiment of the interior door handle. In its final assembly, the conventional door handle consists of at least four major polymer injection molded parts, substructure, trim, and details. When snapped together, those parts are heat staked to the interior door trim panel.

Table 4.3 Extract of exemplary requirements on a conventional automotive interior door handle

Conventional interior door handle design	Simplified list of requirements
	1. Main inside door opening and closing actuation point 2. Comfortable to grab 3. Within one's reach 4. Supportive when getting in and out 5. Pleasant haptics 6. Embedded into interior styling 7. 1st load case - Push/pull force to open/close door e.g., $F_{1y} = 500\,N$ 8. 2nd load case - Push force when leaning from above e.g., $F_{2z} = 500\,N$ 9. 3rd load case - Torque when twisting handle e.g., $M_3 = 3{,}5\,Nm$

The individualization concept for this use case is mostly based on the individual customer attribute *anthropometry*. The anthropometry is unique to each customer and can differ significantly between the 5th and the 95th percentile male or female. Therefore, the individualization concept in this use case aims to enable variably customer-specified door handle reach (compare requirement 3 in Table 4.3), thereby improving ergonomics and comfort for the passenger. The level of shape alteration chosen for this use case is *topology* (Fig. 4.2).

With conventional engineering tools, a parametrization of the conventional CAD door handle design would enable the generation of various variants for differing customer anthropometry. However, during the attempt, limitations of Brep-based parametric associative design in CAD tools nowadays arise. Figure 4.7 shows the parametric architecture of the door handle to be individualized. Parameters addressing the general door handle's position and shape have been implemented into the design. The position is derived from the customer's arm length and the shape of the hand size. By altering these parameters independently, unique part geometry variants can be generated without needing to start the design all over again each time.

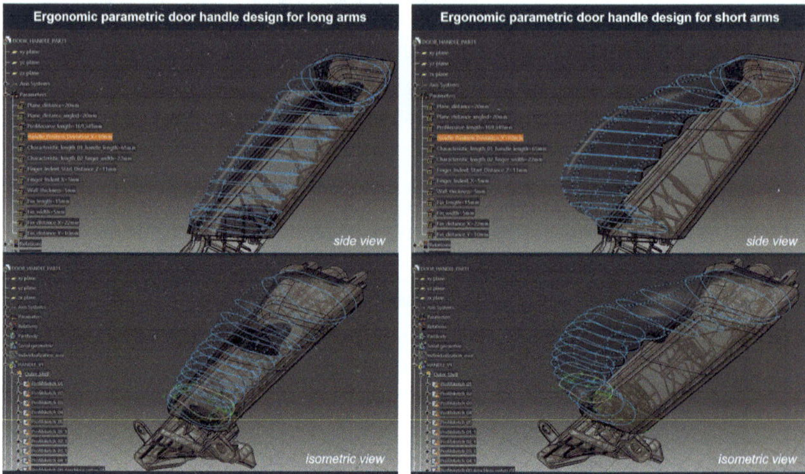

Fig. 4.7 Exemplary boundary representation (Brep) of a conventional (gray) and a parametric individual (black) door handle geometry, designed with conventional CAD tool (CATIA V5, Generative Shape Design)

Regarding the necessity of irregular curved shapes to increase ergonomic comfort, a surface (Brep) based design approach was chosen. With the *multi-section surface* feature of *CATIA V5, Generative Shape Design Workbench*, the outside and inside surface of the parametric door handle are generated based on a multitude of profile sections (Fig. 4.7). At least ten different parameters were necessary to adapt the door handle's shape to the customer's arm length and hand shape. Later, those independent surfaces are joined to a closed, also known as „watertight" Brep, and exported as a printable STL file.

However, even though such a state-of-the-art parametric associative design can reduce variant- or iteration-based engineering design effort, there are several limitations to this approach:

- **Parametric complexity:** Numerous parameters necessary to describe irregular curved shapes
- **Parametric interdependencies:** Limited options to set up relations between mutually dependent parameters
- **Geometric stability:** Spline- or Brep-related instabilities (e.g., continuation errors, cusps, inverted surfaces,...) can lead to collapsing geometric features.
- **Workflow automation:** Even using macros, only limited possibilities to automate the geometry generation process (design engineer needs to be involved)

Due to these limitations, this approach is only suitable for applications with manageable parametric and geometrical complexity. Additionally, the functionality of each geometry needs to be assured with the following operations in CAE tools also challenging to be automated. Furthermore, to minimize development complexity in a mass individualization context, the number of product specification steps customers can fulfill themselves must be maximized. Thus, a different approach to this design challenge is necessary.

As introduced in the fundamentals Sect. 2.1.3 Fig. 2.17, configuration software tools are rather focused on the combinatorics of predefined options, whereas specification tools translate unique customer attributes into unique product designs within the individualization scope. These specification tools are often based on tessellated geometry representations (e.g., STL format). Even though, compared to native CAD files, tessellated file formats contain less geometric information due to their non-associative properties, its triangulated mesh consisting of nodes and triangles enables advanced geometric manipulations with fewer instabilities. This way, geometry representations are mathematically simplified and suitable, e.g., for agile browser-based online specification tools.

Also, a characteristic of specification tools is the purposely limited set of features offered to the customer to interact with the design within the software. Only functions necessary for each step of the specification process are implemented in the tool and active at the right time. The resulting step-by-step specification process is also considered a user journey and, combined with a pleasant user interface design, aims to generate a positive user overall experience.

As a feasibility study regarding this approach's parametric and functional assurance capabilities, a respective specification tool for the individualized door handle application was designed in collaboration with *Trinckle GmbH*. The resulting software is a browser-based tool that could be used by design engineers responsible for generating product variants or directly by customers themselves. However, it is important to clarify the overall specification process prior to the details of the features necessary to individualize the door handle.

The process explained in the following is only focused on the MI context. It might differ strongly in other product individualization settings such as, e.g., special small series production offered by craftspeople (Table 2.7). At least four general process steps can be distinguished when considering product individualization. Even though the customer journey should be set up as interesting and exciting as possible, it is important to remember that product individualization always leads to additional effort for the customers compared to an off-the-shelf purchase. Therefore, the following process steps must be self-explanatory, easy, and time efficient for the customer to execute.

- **Customer involvement:** Engaging customers in the product individualization process in a suitable manner is crucial. Personal customer consultations can be costly and are therefore common for premium or exclusive small-series products. Self-explanatory offline or online specification tools are rather suitable in the MI context. Here, customer involvement means, e.g., marketing and sales pitches, online invitations, and provision of exclusive access or login information to online platforms. Particularly for MI, this step needs to be as straightforward as possible. Otherwise, the customer might lose interest.
- **Product specification:** Once the customer is involved, the product specification process may start. Here, the objective is to gather all customer inputs necessary to fully specify the individual product (compare [Bau07] for detailed methods on product definition). In online specification tools, this is often done step-by-step, leading customers to their final product result. Interactive and responsive input methods, e.g., continuous number sliders vs. type in numeric inputs, while simultaneously updating the product's shape, enhance the specification experience yet might complicate the geometry manipulation and visualization of the product significantly for the system.
- **Manufacturing process:** Besides the crucial requirement that all customer-specified product geometries must be manufactured within the general quality requirements for the production process of choice, additional MI-specific requirements must be considered here. Contrary to mass production, mass individualized parts must be fully traceable to the respective customer. This challenge might demand design measures such as imprinted customization IDs or QR codes on the parts surface or the linkage of related parts (e.g., left and right versions) to minimize sorting efforts during post-processing.
- **Shipment, installation, and use:** This step includes all necessary measures to deliver the individualized product to the customer and ensure its intended use. Here the linkage of the customization ID on the parts with the personal customer data (e.g., name and address) is crucial. Some applications can be individualized while purchasing the product, but others might also be suitable for after-sales individualization. In the latter case, individualized parts can be purchased and self-re-fitted multiple times throughout the vehicle's life cycle. In this case, a proper installation method and respective guidelines need to be provided to ensure the vehicle's functionality and safety.
- **Customer reinvolvement:** The individualization process should not end abruptly with the shipment of the product to the customer. Reinvolvement efforts are necessary to establish long-term customer relations [LRZ06]. Especially for after-sales vehicle individualization applications, the process might be carried out multiple times during the vehicle's life span due to, e.g., changing customer

preferences or vehicle ownership, leading directly back to the beginning of the individualization of the process at step 1.

Kumke et al. (2021) introduce a more detailed description of necessary process steps for MI products enabled by AM [Kum+21]. The objective of the developed door handle specification tool is to cover the *product specification*. Nevertheless, it is crucial to consider requirements from up- and downstream process steps during the development.

Figure 4.8 illustrates the user interface and the main specification steps of the exemplary door handle individualization tool. The tool is structured into five main specification steps. Here, the basic engineering design strategy differs from the conventional approach shown in Fig. 4.7. Instead of adapting the geometry by variation of numeric input parameters for the characteristic surface defining profile dimensions, this approach is based on a morphing feature instead. This feature approximates surfaces to one another. The respective specifications steps are explained in more detail.

Step 1: In this approach, the door handle's surface is approximated to the surface derived from a 3D scan of the customer's hand. This way, the necessary measurements to specify an ergonomic door handle geometry can be reduced significantly. The 3D scan would be generated at a dealership or with advanced smartphone applications at home. Therefore, the specification process in this tool begins with the *upload of an STL-file* (1ST step) generated from a manually operated 3D scanner of a person's hand as the main reference for the upcoming geometry manipulation.

Step 2: The *ergonomic adaption* is specified next. After picking the upper and lower orientation points, the 3D scanned hand is aligned with the door handle base geometry. Number sliders allow for fine translation and rotation adjustments if necessary. Next, the door handles base geometry, mostly in contact with the customer's hand, approximates the 3D hand scan. This morphing feature can be set with variable intensity, resulting in either a close contour or a subtle imprint of the 3D scan on the door handle. This step is not comparable to conventional *Boolean Operations*, in which two volumes can be either fused, subtracted, or intersected [Bri09, 17 ff.]. Here, specific vertices (control points) of one tessellated surface are approximated to the respective vertexes of the target surface. With the forearm deviation adjustment (arm measurement necessary), the door handle is additionally distorted to rather short or long extremities finalizing thereby the ergonomic adaption.

Step 3: The styling of the now ergonomically adjusted door handle geometry is set next. For the *design customization* (3RD step), exemplary style categories are

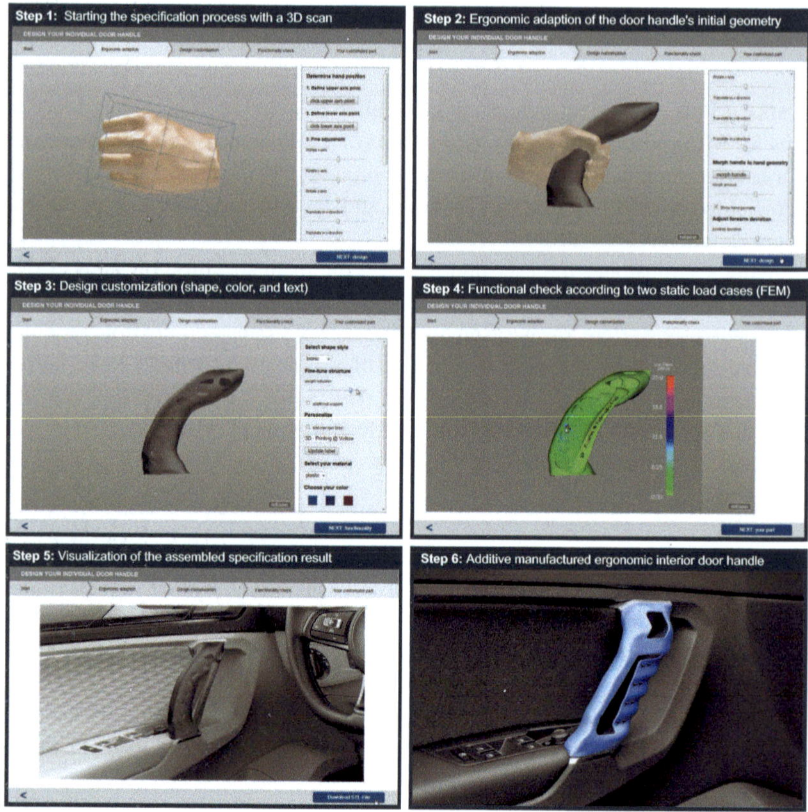

Fig. 4.8 Workflow overview of the main specification steps of a door handle specification tool with focus on ergonomic individualization and its functional assurance. (In cooperation with Trinckle GmbH)

implemented for the customer to choose from (e.g., plain, bionic, dynamic). Different cutout patterns on the door handle's surface represent these styles. In a fully developed scenario, these styles would be professionally pre-developed by the company's design department incorporating the respective design language of the brand. Contrary to conventional product configuration, where static options are just combined, those styles are dynamically adaptable, meaning they can be adjusted and individualized by the customer. Further, the adjustability range made accessible to the customer needs to be set wisely, considering manufacturing restrictions to only allow for feasible designs in each possible setting. In this case, a *weight reduction* slider manipulates the size of the breakthroughs according to the user's preference. An additional personalization option is available to the customer, e.g., by imprinting an own label on the part. With the following choice of color and material, the design customization of the door handle is finalized.

Step 4: In this use case, the individualization options offered to the user significantly impact the door handles functionality. The ergonomic adaption and the design customization can compromise the door handle's structural integrity. This circumstance is an example of function impairing individualization. In applications with high complexity, particularly when considering the uniqueness of the anthropometry of the human body, unpredictable results can be generated with such specification tools. Not all individualization results can be functionally assured beforehand, demanding a functional assurance downstream to the individualization steps. For this purpose, a *functional check* (4TH step) was implemented into this specification tool. Further, realizing a high automation level in this particular step is crucial to managing variant-related assurance efforts. A FEM-based structural simulation is integrated into the specification workflow to ensure the structural integrity of the door handle, considering its static load-bearing requirements (Table 4.3). This way, unpredictable specification results can be directly validated. Structurally weak sections can be mitigated by re-adjusting the weight reduction slider in the previous step and adding additional material to the part. As a validation step, this semi-automated functional assurance should not necessarily concern customers. It might be implemented as a fully automated background validation step or carried out by engineers in the next development stage.

Step 5: In the last specification step, the validated and final geometry of the product is visualized in the respective assembly and exported as an STL file ready to start the manufacturing process (e.g., build job scheduling, data, and machine preparation see Fig. 2.29). In a web-shop setting, the customer would proceed to the general order process, ensuring data for shipment and payment is gathered and the order process completed, concluding the customer journey for this product specification.

This specification tool is one example of various possible applications in the MI context, yet it shows the potentials and limitations of this parametric automated design and validation approach. This tool's schematic back-end system functionalities, necessary to enable the process steps described above, are illustrated in the process overview in Fig. 4.9 and explained in more detail in the following.

The developed system consists of pre-developed elements, a human-machine interface (HMI), and the software back-end to enable this individualization approach. The following descriptions explain those areas in more detail, referring to the process diagram illustrated in Fig. 4.9.

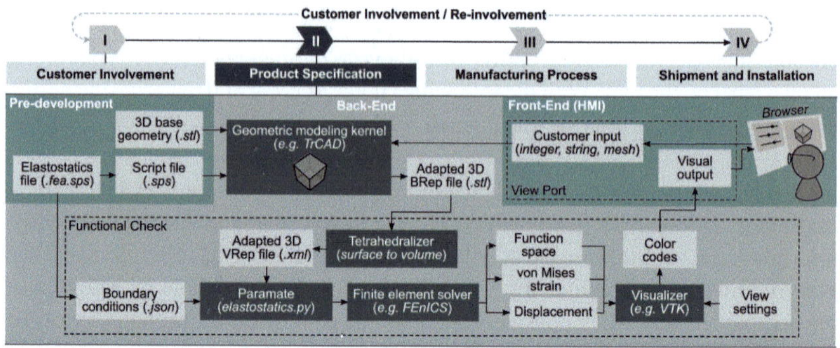

Fig. 4.9 Schematic overview of a exemplary back-end workflow necessary for a specification tool with focus on ergonomic individualization and functional assurance for the mass individualization context. (In cooperation with Trinckle GmbH)

Pre-developed Elements:

Crucial elements for the individualization process must be pre-developed and define the starting point of each individualization journey. For this use case, the 3D base geometry (*stl-file*) of the door handle is needed as the default surface. The script file (*sps-file*) contains all commands necessary for the entire individualization process. Additionally, to define the physical boundary conditions for the downstream structural analysis (e.g., mechanical material properties, load cases), an elastostatics file (*fea.sps-file*) is necessary. With those three elements prepared, all input files are gathered.

Software Front-End (Human Machine Interface):

As this use case is based on a web-based MI approach, the HMI runs in a computer or smart device web browser. A view port is implemented in the respective homepage and is the only interface to the customer. The view port enables customer inputs (e.g., integers, strings, or mesh files such as a 3D scan of the customer's hand) and visualizes the geometric results. The HMI is a crucial element for a straightforward yet exciting customer journey.

Software Back-End:

The software back-end performs all background process steps according to the script file. On the one hand, it contains the geometric modeling kernel, which is necessary to compute all geometric operations and generate the respective representations (e.g., surfaces, volumes) necessary for geometry visualization and export. It receives the customer input necessary to manipulate the default geometry. The back-end contains, on the other hand, all functional operators necessary to carry out the finite-element-analysis (FEM/FEA) of the customer-specified geometry (functional check). First, the *Tetrahedralizer* transforms the customer-individualized tessellated boundary-representation (Brep) into a volume-representation (Vrep) crucial for the material assignment to the part and strain calculation. *Paramate*, the parametrization core of the software developed by Trinckle GmbH, combines the Vrep of the geometry with the boundary conditions defined in the elastostatics input file and provides the data to the finite element solver (FE-solver). The thereby determined outputs, such as stress and displacement values, are prepared for visualization on the part's geometry by the *VTK-visualizer*, and the color codes send as visual output to the HMI-view port, informing the user accordingly.

This process is repeated if changes to the geometry are required. However, the workflow introduced in this subsection suggests one possible approach to enable a fully parameterized ergonomic vehicle component for a mass individualization context. When fully developed, such specification tools bare the potential to reduce variant-based engineering complexity significantly and thus utilize AM manufacturing flexibility with a focus on enabling value propositions for the customer, that are conventionally not achievable.

4.1.4   Discussion and Conclusion

In this section, AM individualization potentials for vehicle components were elaborated, addressing the conflicting objectives of MI market trends and corporate efforts toward product variant reduction. First, general aspects such as possibilities and engineering challenges of mass individualized vehicle components focusing on shape alteration were clarified.

Based on the learnings of three different applications, the first three of the four individualization categories (appeal, functionality, and well-being see Fig. 4.1) and the first three of the four challenges (product specification, engineering design, functional assurance, and brand identity) were elaborated in detail in this section. However, individualized vehicle safety is one remaining category to be analyzed. As the basic idea here is to tailor passive vehicle safety measures to the anthropometry of passengers is similar to the comfort-focused door handle use case, a similar approach could also be applied to safety-related use cases. Nevertheless, functional assurance of safety-relevant vehicle components is highly complex, often demanding multiple validation crash tests to fulfill high-level requirements. Therefore, from an engineering perspective, the mass individualization of vehicle safety aspects is considered too complex and only manageable in the medium- to long-term perspective.

One general aspect of this context, not further discussed in detail in the literature and this section, is the challenge regarding the styling of individualized products. The styling of cars, in particular, strongly depends on brand-specific design languages leading to a maximized brand recognition factor. A conflict with brand identity requirements might occur as soon as a product's appearance is individualized by customer preference. This context adds a new complexity level to future specification tools that must comply with a strict brand identity structure (vehicle proportions, personality, design language, and styling) while enabling sufficient individualization scope for customers. In this case, a suitable compromise is needed combining brand-specific with customer-individual styling elements. One possible approach to mitigate this conflict of objectives is to consider brand-specific styling features (e.g., sharp or smooth angles and patterns) while still enabling the input of individual customer attributes into adaptable part sections. This aspect must be considered in future applications to get clearance from professional styling-focused design departments.

A limitation to the fully parameterized geometries is the robustness of parametric systems. Depending on the complexity of the application, it is challenging to ensure that all intended parameter combinations lead to a stable, manufacturable part geometry. The effort during specification tool development necessary to ensure the robustness of adaptable geometries is still high, particularly when working with native CAD formats instead of tessellated geometry representations.

Even though automated functional assurance by implementing structural simulation steps into specification tools is already a first step towards reducing engineering efforts of individualized parts, a positive simulation result does not fully account for a part's entire clearance according to automotive development processes and quality standards. Additional corporate processes need to be adapted for mass-individualized products. For example, today's one-to-one part identification number must cover several geometrical variants of the same vehicle component in a MI scenario. Moreover, it might be challenging for quality departments to adapt clearance processes to cover a range of possible individualized results while still complying with many quality requirements. Therefore, an adapted PDP is necessary.

Further, the approaches applied in this section were suitable for tangible applications. The applicability of this approach to use cases with higher geometric and functional complexity still needs to be improved due to impairments between different groups of vehicle parts, architecture and computing power of engineering tools, and so on. Nevertheless, shifting the perspective from partial components to the complete vehicle level, exterior individualized vehicle components have been showcased (see Sect. 3.1), and patents already suggest advanced configuration options for entire vehicle interior compositions (e.g., [KBP07]). When scaling AM individualization potential up to the entire vehicle level, individualized vehicle concepts might be conceivable one day.

4.2 Approach Towards AM-Enabled Functional Integration

The fundamentals Sect. 2.3.3 emphasizes that *functional complexity* is one of four major categories of AM-related design principles necessary to obtain a multitude of customer-relevant value propositions. Nonetheless, state-of-the-art AM-focused vehicle concepts do not prioritize functional integration yet (compare Sect. 3.1). Further, as discussed in Sect. 3.4, when applying DfAM potentials to VC holistically and not only part-focused, the resulting interdisciplinary vehicle components show an increased level of part consolidation and thus function integration (compare Table 3.9). However, a lack of methods and tools was identified to manage additional functional complexity (Sect. 2.4.3). This section analyzes a specific automotive application emphasizing a multidisciplinary functional scope to outline challenges and opportunities. This section addresses two of the four identified AM application areas for VC *1. improve driving performance* and *2. optimize package and comfort* (Sect. 3.2.3 Fig. 3.7).

Section 4.2.1 outlines the context of multidisciplinary vehicle development regarding the definition of necessary terms and limitations of state-of-the-art applications. In Sect. 4.2.2 a feasibility study of a multidisciplinary optimized vehicle segment is elaborated in more detail. Respective findings and implications for this work are summarized in Sect. 4.2.3.

4.2.1 Multidisciplinary Vehicle Design Optimization

In the VC context, development activities are often categorized into technical sections, dividing the vehicle as a system into functional scopes (e.g., body, chassis) necessary to manage the specific yet complex development processes (Sect. 2.2.2). Those technical sections constitute disciplines on which the organizational structure of automotive OEMs' development departments can be based. Besides the technical sections of a vehicle, engineering categories, e.g., mechanics, thermo- and aerodynamics, electrics, electronics, and others, are comprehended as engineering disciplines relevant for VC, yet relevant for more than one technical vehicle section.

When using the terms such as *multidisciplinary* or *multiphysics optimization* in this context, it might be unclear if technical sections of the vehicle or engineering disciplines are addressed. For clarification purposes, the following definitions are used in the following:

> **Definition 28:** *Multi-Disciplinary Design Optimization* (MDO) – Several terms, such as *multidomain, multiobjective,* or *multiphysics optimization* are used in the literature referring to certain aspects or alterations of the term *multidisciplinary design optimization.* Nonetheless, in this context, the terms *multidisciplinary optimization* and *multidisciplinarity* refer to the simultaneous consideration of more than one technical vehicle section relevant to VC. Moreover, here the disciplines describe differing functional scopes relevant to VC. However, in the literature, MDO might be used in broader terms, often additionally considering non-technical requirements, e.g., economics, during engineering design [PBD15, p. 6; Sch+17, 4, 17, ff.].

> **Definition 29:** *Multi-Physics Design Optimization* (MPO) – The term *multiphysics design optimization* can be used to describe applications in which technical requirements from more than one physical or engineering category apply (e.g., mechanics, thermo-, aerodynamics, electrics/electronics) [Sie+16, p. 251]. KEYES ET AL. (2012) use the term *mutliphysics simulation* to describe simulation methods that couple multiple physical phenomena [Key+12, p. 1].

Section 2.2.4 emphasizes the relevance of conflicting objectives for VC, as requirements from multiple disciplines need to be considered during engineering design. Particularly when composing the technical package of the vehicle design space, disputations between differing disciplines occur (e.g., drive train and chassis in the vehicle front section). When increasing the functional complexity of parts and taking advantage of AM design freedom, such design space disputations can be mitigated due to minimized partitions and joints. Therefore, weight and assembly time reduction can also be achieved, among other value propositions. The relevance of functional integration as AM design principle for VC has already been identified and implemented into state-of-the-art showcases. One of those use cases is discussed in further detail next.

The project *3i-PRINT*, a cooperation of six companies, elaborated on a functionally integrated vehicle front section on the example of a Volkswagen Caddy (Fig. 4.10) [CSI17]. The design is based on a topology-optimized metal structure focusing on crash-induced load transfer. The resulting, organic-appearing hollow vehicle body section has been complemented with additional functional features such as fluid-containing hollow spaces, cooling fins, air vents for brake cooling, mounting points for chassis components, and others [CSI18].

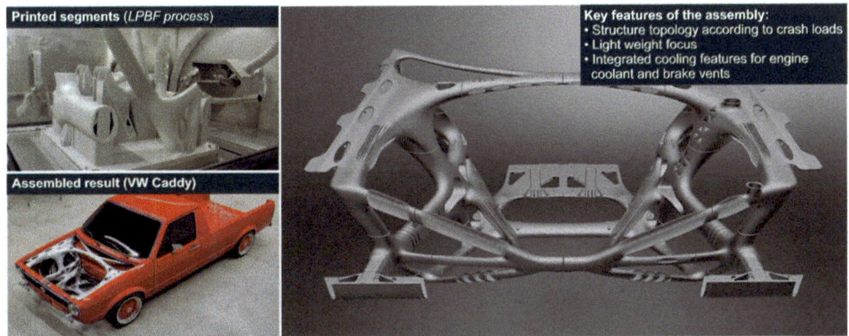

Fig. 4.10 Additively manufactured vehicle structure with focus on functional integration [Alt17; Mor17; CSI18]

Figure 4.10 overviews the virtual model, the additively manufactured segments, and the final result assembled in the vehicle. This approach takes mostly advantage of AM-enabled geometrical, functional, and hierarchical complexity (Table 2.14) considering the manufacturing restrictions of the LPBF process (Fig. 2.30). Even though this approach might not be suitable for mass production due to the low productivity and high costs of LPBF processes, it still challenges conventional vehicle design by functionally integrating and unifying otherwise independent automotive disciplines (e.g., body structure, chassis, thermal management, passive passenger safety measures). However, even though requirements from multiple disciplines have been considered in this welded structure, no specific MDO or MPO method has been applied during the engineering of the concept. Thus, this concept might not be the most suitable compromise between the included automotive disciplines. This approach rather prioritizes the mechanical topology optimization of the body structure as the most important asset of the concept (for more context on *topology optimization*, please compare [Chr+15; Fra18; DPL21; Li+21; Zhu+21; Bar22]). With other words, an optimal load bearing and kinetic energy dissipating structure is designed within the available design space first. The additional functions and features are integrated subsequently, complying with the remaining design space not occupied by the optimized body structure. Therefore, the resulting concept focuses on optimal structural properties yet accepts the sub-optimal performance of the additional functions. For example, even though the topology of the load-bearing struts might be the most suitable for the considered load cases, the thermal cooling features implemented in them might not be placed in an optimal position for the air to stream around and maximize necessary heat transfer. Thus, additional heat dissipation surfaces might be necessary to fulfill the requirements.

In the literature, efforts have been made to develop approaches, methods, and tools to assist MDO and MPO engineering processes systematically. For the DfAM context, at least three of those methodological approaches can be pointed out as suitable for the design of multi-disciplinary AM applications (for a more exhaustive analysis of suitable design methods for functional integration, compare [Kum18, p. 106]).

- **Product Architecture Design:** According to FELDHUSEN AND GROTE (2013), the definition of a *product architecture* is a crucial step in the development of technical systems. In this approach, the product's main function is divided into sub-functions Next, product components are derived necessary to realize those sub-functions Those components are grouped into assemblies, modules, and technical sections, which complete the product's architecture [FG13, p. 257]. Even though this is a conventional approach, the component derivation step is crucial for DfAM applications focusing on functional integration. At this stage, AM-induced design freedom must be considered to determine how many functions can be reasonably fulfilled with a single component (Fig. 4.11).
- **Semantic Network of AM Design Potentials:** The semantic network of AM design potentials established by KUMKE ET AL. (2018) shown in Fig. 2.31, links customer-relevant value propositions with aspects of AM-based functional complexity. The dependency of those aspects from design principles based on form, hierarchical, and material complexity are again clarified. This way, design engineers can be inspired on types of functional elements to be integrated and on additional design principles helpful to do so [Kum+18].
- **Monolithic machine strategy:** This method, introduced by EHRLENSPIEL AND MEERKAMM (2018) is not AM specific, yet it can be applied to optimize existing or develop new product layouts towards integral design or part consolidation. As shown in Fig. 4.12, in the first step, it is suggested to envision all necessary

components of the product as one single monolithic structural element. Static and dynamic components are thereby integrated into the structural element regardless of the necessity of relative movements to each other. This abstraction step shall enable engineers to break free of traditional types of partitions. Subsequently, in the second step, new partitions are only set in crucially necessary areas of the system [EM13, p. 502]. The resulting design might differ from a conventional approach when considering additional AM-related manufacturing freedom during this last step.

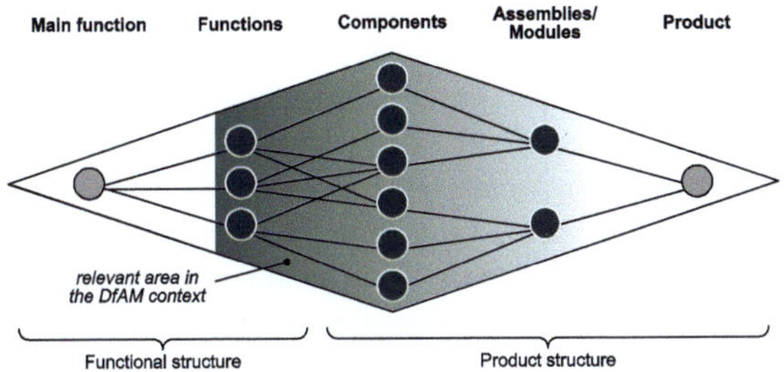

Fig. 4.11 Particularly relevant area of the product architecture definition for DfAM-based functional integration and part consolidation, based on [FG13, p. 257; Kum18, p. 130]

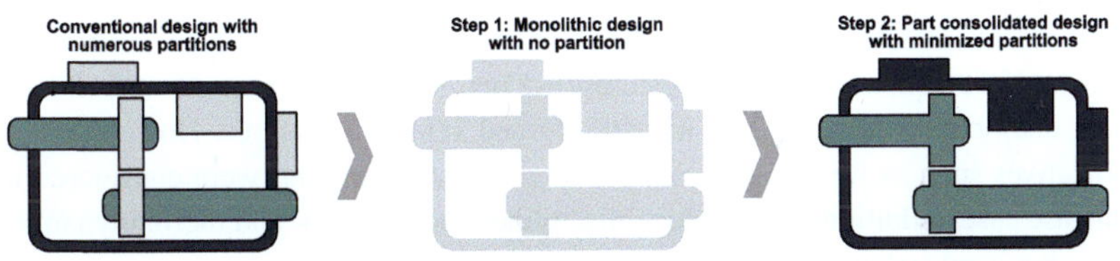

Fig. 4.12 Illustration of the two main steps of the monolithic machine strategy, based on [FG13, p. 257; Kum18, p. 130]

Additionally to methodological design approaches also mathematical optimization methods are being developed and implemented on software-based engineering tools to enable MDO or MPO. Recent advances in the post-processing of tessellated or voxelized geometry representations, e.g., are a first step towards the usability of FEM-based topology optimization results in engineering design. Previously, optimization results obtained with FEM-solvers demanded laborious re-design efforts to generate a CAD-native approximated geometry representation of the optimization result. BARTZ (2022) for instance, introduced a shape optimization approach including an automated transformation of optimization results (voxelized or tessellated) to native-CAD formats (sub-division surfaces convertible to conventional boundary or volume representations) [Bar22; Bar+22]. With such methods, implementing additional functions into topology-optimized shapes is significantly facilitated.

Further, to enable the design optimization of multidisciplinary or multiphysics applications, MDO/MPO software tools focus on coupling up to now independent optimization methods with each other [Sch+17]. Here, the objective is simultaneously optimizing multiple mutually dependent and design-defining parameters. In this context, parameter optimization approaches are often applied to solve complex, conflicting objectives. The interrelations between parameter combinations are clarified through strategic parameter variation, simulation, and numerous iterations. When visualizing the multiple simulation results, so-called *pareto frontiers* [VWZ14; Sch+17] or *sensitivity plots* [PBD15], the possible solution space within the boundary conditions set for the simulation model becomes apparent. This way, design engineers can identify the most viable parameter combination, representing the best compromise of two or more optimization objectives. Several different optimization strategies are being developed in this field of research. For more details on *single-level* (e.g., MDF, NAND, IDF, Single-SAND-NAND, AAO) and *multi-level methods* (e.g., CO, BLISS, DIVE,) compare [Bal+12, p. 626; Key+12, 28 ff.].

Examples for common commercially available tools used for multiphysics simulation are COMSOL Multiphysics®, Altair SimLab®, Dassault Systèmes SIMULIA™, or Siemens Simcenter®. Here, most engineering fields, such as structural mechanics, electromagnetics, fluid dynamics, acoustics, and thermodynamics, are implemented and can be coupled during optimization. VELEA ET AL. (2014), e.g., elaborate on an MDO of fiber-reinforced vehicle structures. However, only objectives such as weight, material costs, and design stiffness were considered in this use case. Additional automotive disciplines such as aero- and thermodynamics have not been considered [VWZ14].

Moreover, GEBHARDT ET AL. (2018) suggest an approach that focuses on determining principle stress lines inside a certain design space for reinforcement purposes (Fig. 4.13 top). This approach can be considered a simplified strategy to comprehend a given design space's load characteristics and visualize the most relevant load paths. BIEDERMANN ET AL. (2022) follow a related idea, yet focus on fluid instead of load paths (Fig. 4.13 bottom). Here, the automated generation of flow channels for complex hydraulic manifolds is targeted [BBM21; BBM22]. Even though those methods are not fully developed yet, similar approaches for many physical engineering categories might become helpful for early design stage MDOs and some day for VC.

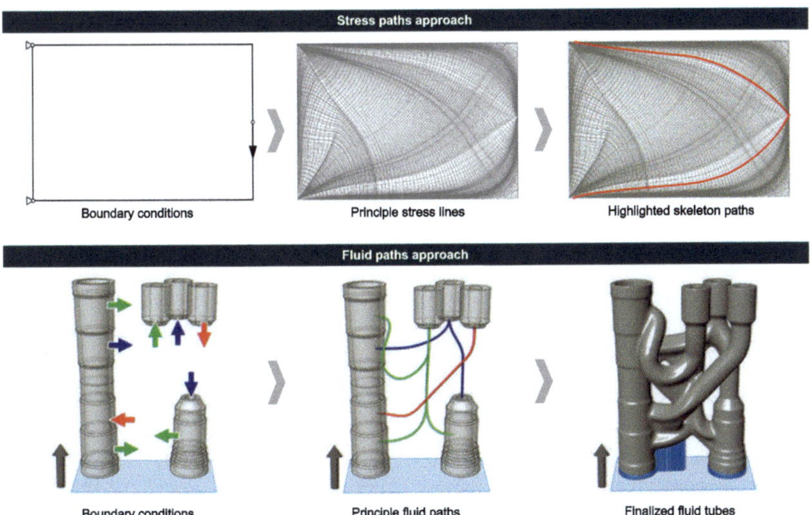

Fig. 4.13 Determination of structural stress paths (top) and fluid flow paths (bottom) for optimization purposes in predefined design space [GTV18, p. 795; BBM22, p. 2]

Concluding, MDO and MPO approaches are mathematically limited to target functions that prioritize the multitude of optimization targets in relation to each other. Further state-of-the-art approaches are limited to parameterizable designs, in which the available design space is already prioritized and pre-sectioned between the respective disciplines or physical engineering categories. This way, disciplinary design approaches are encouraged, and significant functional integration potential on the complete vehicle level still needs to be leveraged. Up to now, the focus has been on fine-tuning already available designs towards local optima and not enabling multidisciplinary early-stage optimization on the vehicle concept level towards a global compromise. Further, as additional AM-related efforts and costs can be justified with better-performing designs, the necessity is identified to analyze the opportunities and limitations of early-stage MDO/MPO approaches for VC in more detail. Hence, a respective use case is elaborated accordingly in the following subsection.

4.2.2   Development of a Multiphysics Optimized Vehicle Segment

This subsection systematically elaborates a use case to explore the impact of DfAM-enabled functional integration and multidisciplinarity on vehicle conception. As often stressed in the literature, DfAM approaches motivate design engineers to think outside the box and break loose from conventional development patterns and manufacturing restrictions [BGH05; YZ15; Tho+16; Sch+19; Gib+21, p. 567]. Thus, and with the limitations of state-of-the-art approaches clarified in the subsection above in mind, this analysis aims to approach the conception of a vehicle without strict compliance to conventionally separated engineering fields, organizational disciplines, technical sections, or physics areas. Additionally, even though it is important to determine local optima for each relevant function, the best holistic compromise for the vehicle is the main priority. Nevertheless, for complexity management purposes for this thesis, the engineering problem is reduced to just a representative section of the vehicle. The vehicle section should have a manageable functional complexity, yet still be large enough to bare relevance to VC.

The respective use case is based on a fuel cell-powered electric drive vehicle concept. Figure 4.14 shows the conventional unibody shell design of the identified vehicle area in focus of this subsection. In this rear section of the main underbody, multiple functionalities of differing engineering fields need to be realized. Besides the partial vehicle underbody, this segment also includes the lower side panel. Requirements from four technical sections (body system, drive train, chassis, electric/electronics) apply to this vehicle cutout making it particularly interesting for interdisciplinary analysis.

Besides the technical vehicle sections, diverse physics categories are also relevant for this use case. In Fig. 4.14, the main multiphysics interfaces of the underbody segment are also visualized. For this context, the physics categories are differentiated according to the type of energy transfer relevant to the functional requirements. The first category *energy transfer* in terms of mechanical loads is derived from chassis- and crash-induced load cases. *Mass transfer*, the second physics category relevant to this vehicle section, is mostly important for the fuel-cell system regarding hydrogen and air supply and exhaust water. Here, also *heat transfer* occurs due to the exothermic nature of the electrochemical redox reactions in the fuel cell. The fourth category focuses on the *transfer of electric current*. Numerous wires throughout the vehicle are necessary to conduct electricity between energy sources, sensors, actuators, and processing units. In this vehicle section, particularly the high voltage interfaces to and from the battery are considered, as these cables have high cross sections and therefore need to be considered in the technical package

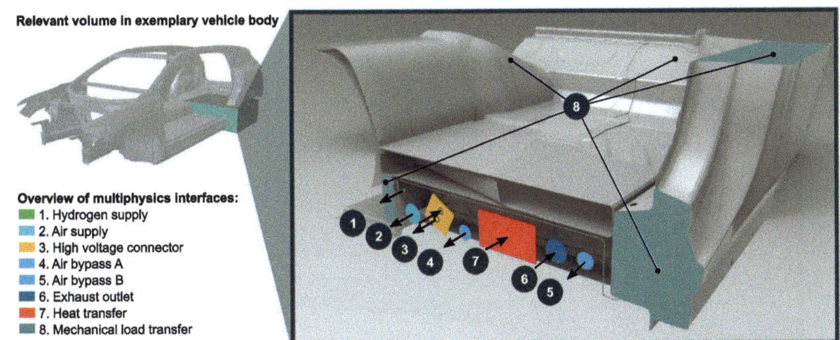

Fig. 4.14 Conventional vehicle underbody segment identified for multidisciplinary design analysis including overview of multiphysics interfaces relevant for this vehicle concept

of the vehicle concept. Table 4.4 summarizes those four physical areas and their assigned high-level requirements. Additionally, Table 4.4 also indicates the respective technical vehicle sections to which the requirements belong. A comprehensive list of requirements for an entire vehicle segment might be far more exhaustive, as shown in Table 4.4. Additional requirements such as waterproofing of passenger compartment and acoustic insulation also apply here, yet are not prioritized for this high-level analysis.

As stressed in the fundamentals Sect. 2.3 Table 2.11, AM-related design freedom enables manufacturing geometries closer to computationally optimized ones in comparison to conventional manufacturing processes. With this in mind, a two-step engineering strategy elaborates on this multiphysics analysis. The objective of the first step is to obtain local optima to fulfill the requirements for each physical area individually, considering the boundary conditions of the available design space. This way, each vehicle's physical area and technical section are taken into account, unaffected by each other. Additionally, the most relevant design space areas for each function become apparent. Those local optima are only superpositioned in the second step, potentially revealing design space conflicts between physical areas. Additional design complexity might be necessary to solve those conflict areas. This way, a global compromise considering local optima is targeted. In the following, the differing approaches to obtain local optima are discussed separately for each relevant physical area of Table 4.4.

Table 4.4 Exemplary functional requirements of a vehicle underbody segment according to physical areas and technical vehicle sections for a multiphysics analysis. (*B = body system, C = chassis, D = drive train, E = electrics/electronics*, compare ESM App. A.6 for quantified requirements)

Physical Area / Requirement	B	C	D	E
1. Mechanical load transfer				
1.1. Absorption of side crash impact	•			
1.2. Absorption of rear crash impact	•			
1.3. Support dynamic driving forces induced by subframe mounts (six worst load cases)		•		
1.4. Weight force of technical components and passengers	•	•	•	
2. Mass transfer				
2.1. Hydrogen supply			•	
2.2. Air supply			•	
2.3. Air bypass I			•	
2.4. Air bypass II			•	
2.5. Exhaust			•	
3. Heat transfer				
3.1. Heat flux			•	
4. Electric current transfer				
4.1. High voltage interface I				•
4.2. High voltage interface II				•

An abstraction of the design space, as shown in Fig. 4.15, is necessary to do so. In this step, the design space is volumetrically partitioned into design and non-design areas. The design area is limited by the outer vehicle contour, interior space, and components considered in the technical package (Fig. 4.15). The non-design areas consist of mechanical, functional, thermal, or electrical interfaces that are predetermined in position and dimension based on the technical package. With the resulting characterized design space and the quantification of the relevant requirements, the elaboration of local optima is prepared.

The main focus of the following analysis for each physics area is not to directly obtain the optimized and final geometry for a certain function. The aim is rather to identify the most relevant areas of the available design space for a specific function. The detailed geometry should only be in focus after the second step, the superposition of the local optima. As available multiphysics software tools focus on the parameter optimization of parameterizable geometries, those tools cannot be considered in this early stage of design space analysis yet. Therefore, alternative approaches are considered in the next subsections.

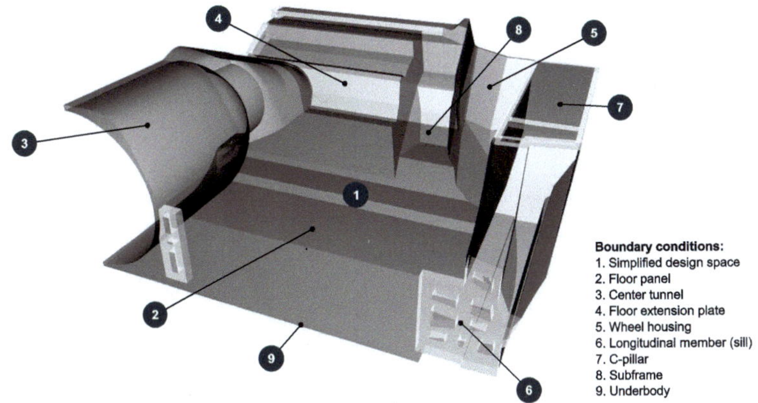

Fig. 4.15 Overview of volumetric design space available for multiphysics analysis in the vehicle underbody section

Mechanical Load Transfer

It is important to interconnect the already defined mechanical interfaces to determine the most relevant design space areas for the mechanical load transfer in this vehicle segment (Fig. 4.15). Without knowing directions and the amount of loads to be transferred between each mechanical interface, the material and hence the volume necessary to do so cannot be estimated. Their simple interconnection is, therefore, not significant. Nevertheless, computationally simple methods are preferred at this early design stage to save effort and time as alterations will be necessary. The application of the *principle stress line* (PSL) approach by GEBHARDT ET AL. (2018) mentioned above already results in significant interconnections, so-called stress lines based on the connection of principle stress vectors on support points inside the design space. Compare ESM App. A.6 Tab. A.7 for quantified load cases considered for this use case. Unfortunately, when transferring the PSL method from 2D to 3D space, the resulting stress lines or load paths can seem chaotic. It remains challenging to estimate the volume of the design space necessary to transfer or bear the respective mechanical loads properly (compare Fig. 4.16).

The computationally intensive topology optimization method is applied next to obtain a volumetric result of load transfer relevant areas. In the FEM-based topology optimization, the material-filled design space is incrementally reduced until just enough material is left to comply with the simulation model's boundary conditions. With multiple iterations, only the most relevant load-bearing finite elements

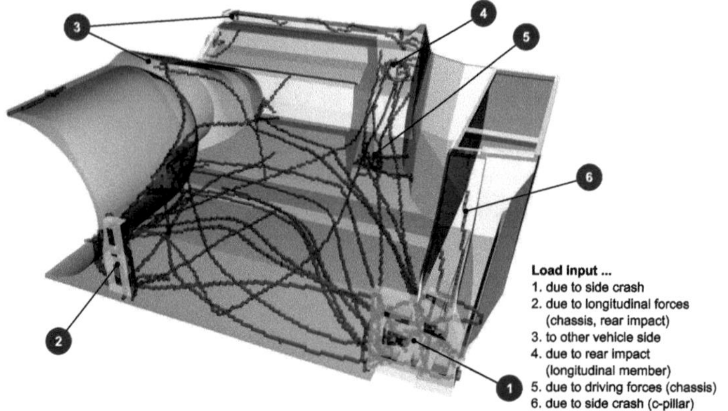

Fig. 4.16 Principle stress lines in vehicle segment design space according to load conditions on mechanical interfaces based on [GTV18]. (For exemplary load cases compare ESM App. A.6 Tab. A.7)

remain in the design space. Besides the abstracted design space and the mechanical interfaces (non-design) (Fig. 4.17), boundary conditions such as effective loads, bearings, or rigid elements are necessary to complete the simulation model. With the definition of constraints, such as maximum allowed displacement values for certain control points in the structure, or maximum allowed stresses in the structure, the optimization process toward weight minimization is carried out iteratively. The design iteration which complies best with the boundary conditions and constraints with minimum weight is chosen for detail design. The optimization for this use case was carried out considering the exemplary load cases listed in ESM App. A.6 Tab. A.7, the FEM-deck defining the simulation model was set up with the FEM pre-processor software tool ANSA® (ESM App. A.6 Fig. A.2), and the topology optimization conducted with the commercial optimization environment LEOPARD of Volkswagen AG [Fie16; Fra18; Bar22]. Unlike conventional topology optimization tools, LEOPARD is able to post-process voxelized rough FEM geometries in a density-based shape and surface optimization step, also enabling CAD-export of the result [Bar+22].

At this point, it is crucial to emphasize that topology optimization processes are mostly suitable for linear-elastic structural analysis. To consider the dissipation of crash energy by structural deformation in topology optimization (crashworthiness) is not yet possible due to the nonlinearities concurring in those applications [Ort+21, p. 1]. Alternative methods such as the *graph and heuristic topology optimization*

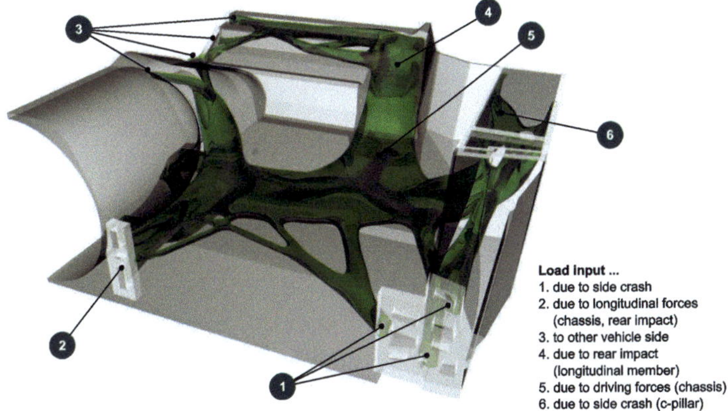

Fig. 4.17 Optimized structure topology in design space of vehicle segment according to load conditions in mechanical interfaces. (For exemplary load cases compare ESM App. A.6 Tab. A.7)

suggested by ORTMANN ET AL. (2021) are being developed to consider nonlinearities. However, up to now, this method is only applicable to two-dimensional applications (e.g., cross-sections of extrusion profiles). Therefore, the maximum occurring forces in two crash scenarios (side pole and rear crash) have been considered representative loads for this early-stage static analysis. Nonetheless, crumple zones should already be considered during the development of the technical package.

Concluding, even though structural topology optimization depends strongly on material properties and boundary conditions of the simulation model, which might be only vague during early-stage concept development, the method can be considered a useful tool for determining relevant design space areas for mechanical load transfer. Additionally, those mechanically sensitive design space areas are this way determined without considering conventional manufacturing restrictions, thereby taking an important aspect of DfAM into account.

Mass Transfer

Similar to mechanical load transfer also to determine the most relevant design space for mass transfer, a method is targeted to identify the most significant interconnection paths between media inlet and outlet ports (Fig. 4.14). This use case includes mass transferring inlet and outlet ports for hydrogen, air, and exhaust water due to the fuel cell drive train of this vehicle concept. Comparing the exemplary input flow rates, densities, and pressure levels from hydrogen (e.g., <1,5 g/s) and air (e.g., <60 g/s) summarized in ESM App. A.6 Tab. A.7 it becomes clear that the air supply is the most package restrictive mass transferring duct for this vehicle segment. Therefore, this analysis focuses on the air supply. Nonetheless, the elaborations discussed below also relate to the general conduction of gases or fluids.

As for structures, also in computation fluid dynamics (CFD) topology optimization is applied to gas/fluid conducting parts. However, to identify the most relevant design space for the air supply, a sensitivity analysis can already be helpful to avoid computationally intense process steps in the early design stage. Sensitivity analyses are often used as pre-simulation steps to verify if the CFD model is set up realistically. For more context on sensitivity analysis and shape optimization of fluid structures consult, e.g., the elaborations of NAJIAN ASL (2019) [NA19, 4 ff].

Figure 4.18 illustrates on the left the heat map results of a CFD sensitivity analysis carried out in OpenFOAM® (open source field operation and manipulation). It can be observed that the dark red colored areas close to the air inlet port fed by the compressor have the highest sensitivity for mass transport in both cross sections. On the right, Fig. 4.18 visualizes an isometric surface including 90 % of the most sensitive finite elements to air mass transfer. Here, it is important to emphasize that those isometric surfaces can not be considered part geometry. They rather indicate

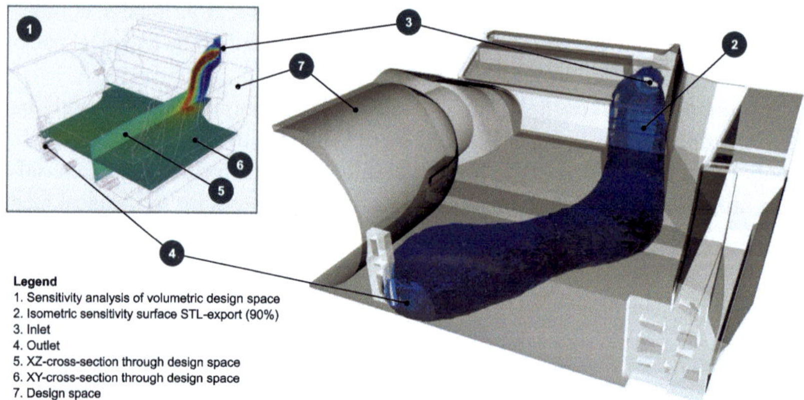

Fig. 4.18 Overview of a CFD sensitivity analysis of design space (*left*) of a vehicle segment according to air inlet and outlet ports and the resulting isometric 90 % sensitivity surface (*right*) (Sensitivity simulation carried out with OpenFOAM®)

the importance of the respective design space area for a particular function. Also, no optimization process has been carried out yet. Only the initial base conditions of the simulation model are considered. However, even though a topology and shape optimization of this use case might result in a different geometry at the end, with this step, the most relevant mass transfer area of the design space according to the inlet and outlet port's position is already identified with short computational time (< 5 min.).

The sensitivity approach might not lead to significant and distinguishable results in the case of multiple mass-transferring structures in the same CFD simulation model. In this case, a combination of individual sensitivity analysis and the approach introduced by BIEDERMANN ET AL. (2022) referred to in Fig. 4.13 might bear further potential [BBM22].

Electric Current Transfer

Next, the high-voltage interfaces need to be interconnected to realize the electrical requirements. This process is often done manually, considering the design space left by other engineering disciplines. As their constant cross-section also predefines the shape of electric cables, there is no focus on shape optimization set here. Nevertheless, the shortest interface connections are targeted to minimize transmission losses and ensure electrical efficiency. CAD software has implemented tools to assist with such design tasks. For example, the feature *path finder* in CATIA® V5 is designed to find possible interface connections inside predefined design boundaries automatically. As shown in Fig. 4.19, even though only mixed results could be achieved and a random behavior was observed, the feature is capable of generating reasonable

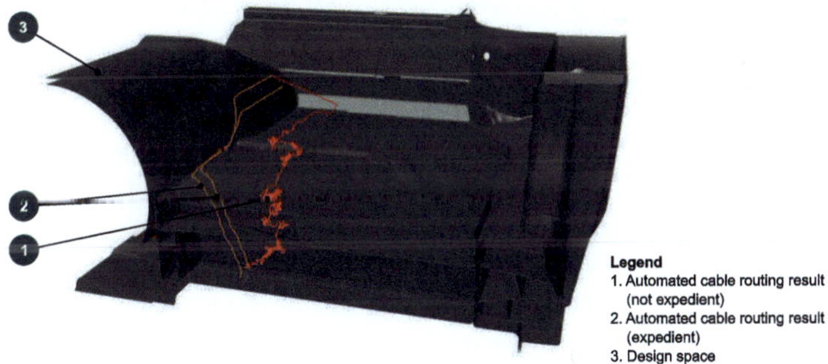

Fig. 4.19 Results of automated cable routing connecting electrical interfaces of a vehicle segment (CATIA® V5 pathfinder feature)

paths avoiding collision with the design space with short computing time ($<$ 1 min.). Nonetheless, it is not yet possible to generate multiple collision-free paths simultaneously that keep a minimum distance from each other. Those features could be developed further toward path length optimization for generating multiple simultaneous routings. Also, a combination with the approach suggested by BIEDERMANN ET AL. (2022) could also lead to collision-free yet efficient routing in complex multiphysics environments [BBM22].

Heat Transfer

Considering heat transfer as the last physics area of this use case shows that research also focuses on topology optimization to determine local optima for coupled thermal-fluid systems. BIEDERMANN ET AL. (2022), e.g., introduce an adjoint approach to this context [QD16]. Unfortunately, those approaches are limited to two-dimensional applications and thus not applicable to this use case. However, most of the thermal geometrical design problems in the automotive sector can be reduced to the conflicting objectives of maximizing heat exchange surface by minimizing pressure loss of coolant or airflow. Additional parts (e.g., radiators) are necessary to achieve those surface maximized geometries conventionally. Considering the AM-related geometrical complexity, the potential arises to integrate surface-maximized geometries hence radiators, directly into in structural components of the vehicle. This approach is also considered in the benchmark example shown in Fig. 4.10 and not further elaborated for this use case.

Superposition local optima to achieve global compromise

With the determined local optima for each physical area of the design problem, the second step of this two-step engineering strategy follows. Here, the predetermined local optima are superposed to reveal conflicting design space areas. If conflicts occur, additional geometric or functional complexity might be necessary to develop the most suitable global compromise based on local optima. For this use case, this step is conducted with the superposition of the structural topology optimization with the CFD sensitivity analysis for air transfer elaborated above. Based on the reference vehicle's package, the air supply (compressor) is closely positioned to the rear longitudinal member of the body. Thus, the sensitivity analysis shares structurally important design space with the topology optimization result. Figure 4.20 illustrates this superposition displaying major conflict zones.

In conventional vehicle conception, technical disciplines are prioritized, thus hindering such multiphysics conflicts from happening yet not efficiently utilizing the available design space. At this stage, it is up to the design engineers to decide

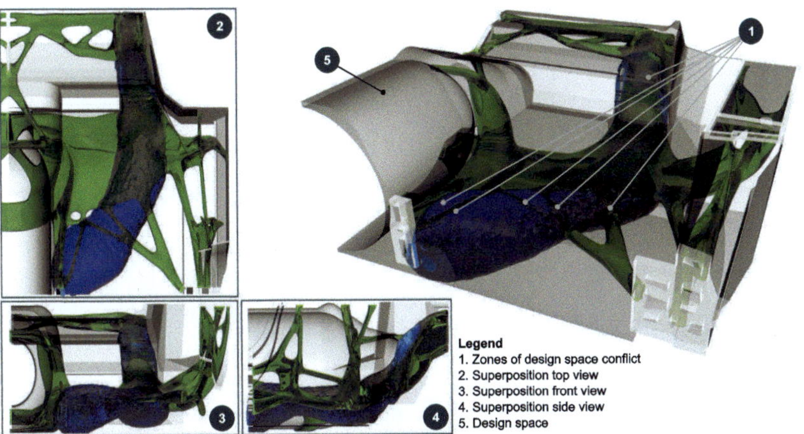

Fig. 4.20 Resulting conflicts in design space of vehicle segment according to superposition of load and fluid relevant volumes

how to cope with the uncovered conflicts, as there are no commercially available software tools. There are at least two differing strategies to proceed with this design problem. On the one hand, it would be possible to partition and restrict the available design space for some physical areas and alleviate it for others. In this use case, e.g., this would mean subtracting the safety-relevant body structure from the available design space for the air transfer, thereby accepting increased pressure losses due to prolonged ducts. Strategies on how to cope with design space conflicts are further elaborated in Sect. 6.3.

On the other hand, the additional DfAM-based geometrical, functional, or hierarchical complexity can be applied to generate functionally enhanced multiphysics geometries. Particularly consolidation, functional integration, or functional extension of parts might be in focus. Figure 4.21 illustrates how geometrical and hierarchical complexity might be useful in an exemplary design space disputation. Such geometries as shown in Fig. 4.21 strive for the best compromise between conflicting local optima demanding design freedom not granted with most conventional manufacturing processes. Here, only one example is illustrated, in which DfAM-enabled functional integration has substantial impact on vehicle conception.

The integration approach chosen to solve the design space conflict in Fig. 4.21 demands compromises from both functions, mechanically efficient load transfer and dynamically efficient fluid transfer. These compromises are illustrated in further detail in Fig. 4.22. Pos. 3, e.g., indicates that even though a strut has been

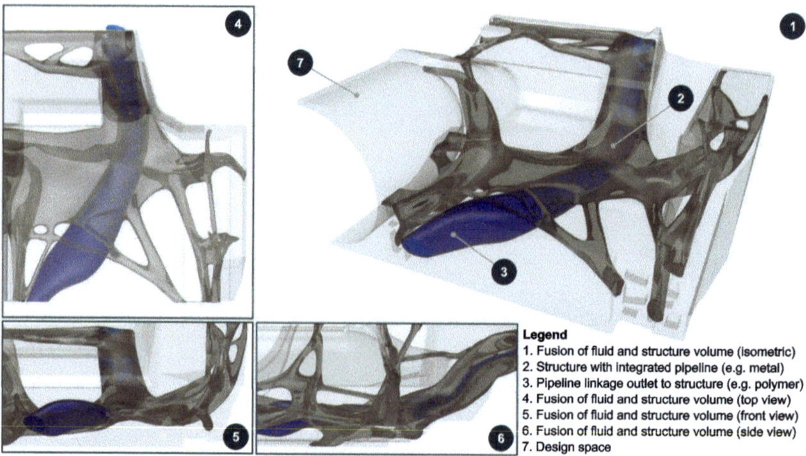

Fig. 4.21 Exemplary integration of fluid-relevant volume into structure considering DfAM design principles towards minimal compromise to achieve global optimum for multiphysics vehicle segment (Manual fusion of OpenFOAM® and LEOPARD® results, designed with CATIA® V5 IMA, GSD environment, and Rhinoceros® Grasshopper)

integrated into the air supply, this same strut has been aerodynamically streamlined to reduce drag, resulting in additional structural weight. Pos. 4 points to a significant strut that has been split into two curved halves and implemented into the metal section of the air supply housing to avoid additional pressure losses. The air supply housing has been partially integrated into the metal structure, stabilizing the split curved struts further additionally adding stiffness yet also weight to the structure, as shown in Pos. 5. The integration of the primary section of the air supply, a significant hollow volume, into a structure leads to diminished stiffness properties endangering its structural integrity under load. Therefore, stiffening struts have been randomly yet parametrically distributed inside the integrated air supply section (Pos. 6). These struts feature an aerodynamically advantageous elliptical cross-section that is aligned to the central spine and thus to the flow direction of the air supply, minimizing induced pressure losses. However, although each feature mentioned above impacts the opposing functional optimum negatively, the resulting global compromise between these two functions only deviates slightly from the respective local optima. With all the limitations of this analysis in mind, this concept consumes, at this rough stage, less than 20 % of the design space consumed by the conventional sheet metal solution. Consequently, a maximum of 80 % additional design space is now available to realize and optimize all remaining functions (e.g., improving passenger ergonomics in the interior, lowering the center of gravity, and integrating additional functions in the same package).

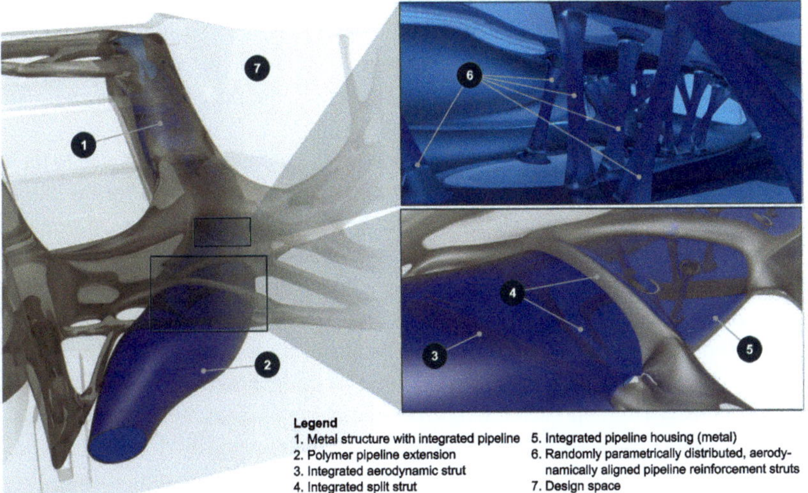

Fig. 4.22 Design details of exemplary integration of fluid-relevant volume into structure considering DfAM design principles (Manual fusion of OpenFOAM® and LEOPARD® results, designed with CATIA® V5 IMA, GSD environment, and Rhinoceros® Grasshopper)

4.2.3 Summary and Discussion

Section 4.2 analyses the potential impact and challenges of DfAM-enabled functional integration on use cases relevant for VC. Here, the two of four AM application areas *1. improve driving performance* and *2. optimize package and comfort* are in focus. Further, the often ambiguously used terms MDO and MPO are differentiated from each other for this context (Sect. 4.2.1). The analysis of the benchmark project *3i-PRINT* shows that DfAM-enabled functional integration on VC level can be considered state-of-the-art. Nevertheless, it becomes clear, that additional systematic scientific approaches are necessary to leverage AM design freedom holistically to not rely on prioritization and optimization of a single technical vehicle section. Additionally, contemporary research work, relevant for early stage design space exploration, has been pointed out.

Based on those aspects a concept study on a multiphysics use case relevant for VC was conducted next. A segment of a fuel cell powered vehicle concept was analyzed in detail. The objective was to identify local optima for individual functions independently from conventional technical vehicle sections or manufacturing processes. Next, conflicting design space areas between two conflicting functions (e.g., crash load and air supply) were identified with the superposition of both local optima. DfAM design freedom was subsequently applied to those conflicting areas to obtain a global compromise between individually optimized functions without prioritization of one function or another.

Even though MDO/MPO software tools are commercially available, due to their restriction to parameter optimization, no suitable tool was identified for the early-stage MPO in this use case. Nevertheless, the elaborated concept independently considers contemporary advances in the computational optimization of each relevant physical area of the use case. The vehicle segment could not be developed in detail with state-of-the-art engineering tools and within the resources of this thesis. Nonetheless, the result achieved demands of up to 80 % less design space, promising less material usage by considering local optima for each specific function simultaneously. This approach is expected to improve the package (e.g., by smaller profile cross sections and shorter air supply ducts) and increase driving performance (e.g., by optimized body stiffness, lower center of gravity, and minimized pressure losses in air supply). Further, significant design space for optimizing or integrating remaining functions has been liberated.

Moreover, this use case implies limitations necessary to mention. The technical package of the vehicle concept considered in this use case is a result of conventional automotive engineering strategy. A different package composition might result when considering DfAM potentials also during the development of the vehicle's technical package. Even though this analysis focuses on a fuel-cell-specific functional scope, the principle methodological approach is meant to be applicable throughout the vehicle independently of drive train type and physical area. DfAM enabled design freedom encourages design engineers to strive for technically optimal shapes and geometries. Considering the high costs related to AM technologies, only geometrically and functionally complex sections of the vehicle, e.g. conflicting areas in design space, might be economically feasible when additively manufactured. The remaining optimized geometries might also be conventionally manufactured if possible. In dependence of the AM technology of choice, respective manufacturing restrictions need to be subsequently implemented into the design during design detailing phase. Those restrictions have the potential to restrict the design's performance further. Not to neglect, this approach demands nowadays separated automotive disciplines, technical sections, and departments to work closely together. This process bears the potential to increase development complexity from an organizational point of view. Lastly, even though functional integration and part consolidation bear the potential to optimize the vehicle's performance and package, it is important to point out, that ease and cost of repair can be worsened as small parts might not be easily replaced in monolithic design architecture. Though, this is a independent debate in which also the reduced number of potential failure sources such as joints and seals in monolithic designs need to be taken into consideration.

Concluding, with this exemplary design study this section emphasizes, that AM-related design freedom can have substantial impact on a vehicle's conception from a functional integration and optimization perspective. Even though DfAM increases feasibility of computationally optimized geometries, also challenges and limitations in necessary engineering methods and computational software tools were identified in this context, thereby impeding the holistic utilization of AM-related design potentials. The methodological generalization of this approach and its impact on design processes is further discussed in Chap. 6 Sect. 6.3.

Leverage Effects of AM Design Potentials on the Complete Vehicle Level

5

Subsequently to the analysis of the potential impact of AM design freedom on specific and tangible automotive applications in Chap. 4, this chapter raises the focus from specific applications up to the complete vehicle level (Sect. 1.4). This step outlines the potential impacts of AM-related design freedom on VC holistically.

The two engineering challenges *variant-* and *functional-related* engineering complexity, identified in the key application areas of AM potentials (Fig. 3.7), have been addressed in Chap. 4. However, for automotive OEMs to assess the disruptive potential of AM technologies, it is crucial to outline the holistic impact of new design potentials on the complete vehicle level in a qualitative and, if possible, quantitative manner. Considering the engineering complexity of modern vehicles, it is challenging to quantify the holistic impact of DfAM on VC within the scope of a thesis. Consequently, the tangible application of one AM-related design principle *AM-based lightweight design* was chosen for an exemplary technical up-scaling analysis from part to complete vehicle level. For this purpose, Sect. 5.1 introduces a technical up-scaling approach to obtain a quantified estimation of the obtainable AM-related lightweight benefit to a respective vehicle.

Supplementary Information The online version contains supplementary material available at https://doi.org/10.1007/978-3-658-49677-7_5.

> **Definition 30:** *Technical Up-Scaling* (TUS) – In this thesis, TUS describes the process of projecting a technical principle that has already been applied on the part level to the subordinated, more complex levels of the product's architecture. TUS simulates the application of a specific technical principle throughout the entire technical system to estimate its potential effects on the product or, in this case, on the vehicle holistically.

In the second part of this chapter (Sect. 5.2) AM-inspired alternative approaches to conventional vehicle body design (Table 2.9) are elaborated. Here, DfAM-based design principles are applied to the vehicle body as the fundamental technical section of VC, generating differing approaches to additive manufactured vehicle body structures. Further, those approaches are characterized and compared qualitatively.

5.1   Assessment of the Impact of AM-Based Lightweight Design on Vehicle's Track Performance

Besides advancements in material and manufacturing technologies such as increased usage of laminated fiber composites [BS13; BF18], progress in simulation methods, particularly in topology optimization, is a crucial element of lightweight design. Depending on the boundary conditions, topology-optimized geometries often consist of complex bionic shapes that might be challenging to manufacture and are therefore restrained to process-related manufacturing restrictions [Fra18]. Combining topology optimization methods and manufacturing processes with high degrees of design freedom, such as investment casting or AM, provides enhanced potential for lightweight design. This section unveils and quantifies the potential impact of the combination of lightweight design through topology optimization and AM-related design freedom on vehicle's performance and is based on the elaborations of TSCHORN, FUCHS, AND VIETOR (2021) [TFV21]. The primary customer-relevant value proposition chosen for this analysis is the *track performance* of a sports car, as this property is one of the main drivers of lightweight design in this vehicle segment. Furthermore, this analysis proposes an *technical up-scaling* (TUS) approach based on identifying part relations between state-of-the-art use cases and the exemplary vehicle itself.

5.1.1   Methodological Approach

Following HENNING AND MOELLER (2011), and KLEIN AND GÄNSICKE (2019), several lightweight design strategies can be differentiated (Fig. 5.1) [KG19; HM11]. These should be considered collectively to achieve maximized weight savings during the vehicle development process. The following analysis focuses on the geometry-based lightweight design approach only. The main objective of *geometry-based lightweight design* is to minimize the mass of load-bearing structures according to their requirements by optimizing the load distribution and shape of parts or assemblies.

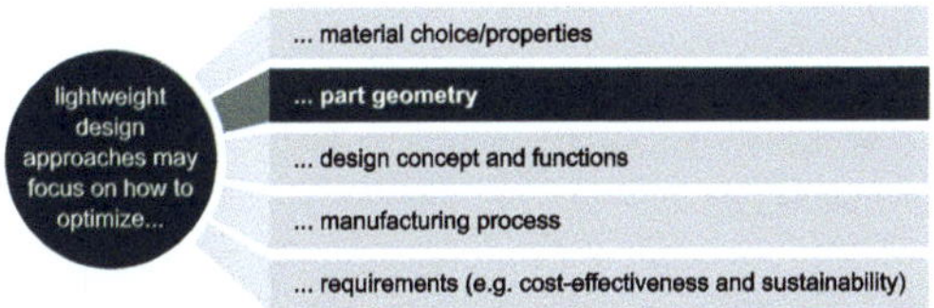

Fig. 5.1 Overview of lightweight design strategies with focus on part geometry adapted from [KG19; HM11; TFV21]

The developed methodological four step approach for this analysis is outlined in Fig. 5.2. The basis of this work is formed on data collected from exemplary applications and use cases of additively manufactured state-of-the-art vehicle components with a focus on lightweight design. According to the majority of applications published in the AM lightweight design context, this analysis focuses on metal components only. To enable a lightweight-focused analysis of the impact of AM and topology optimization on a vehicle in its entirety, a sports car (Bugatti Chiron) was chosen as a reference vehicle. Part relations between use cases and reference vehicle were identified and potential weight savings summed up. The impact of those savings on vehicle's track performance was quantitatively determined by a vehicle dynamics simulation. In the following the individual process steps are explained in more detail.

Step 1 and 2: Collection of AM-related Lightweight Use Cases and Vehicle Data

The necessary data is compiled from published AM lightweight-focused vehicle applications (database 1 in Fig. 5.2). These components are partially known to the public through literature and, to some degree, communicated internally in a corporate environment. Parts with an insufficient information or development maturity are not further considered. The collected data is subjectively evaluated and filtered taking the following key attributes into account (Table 5.1).

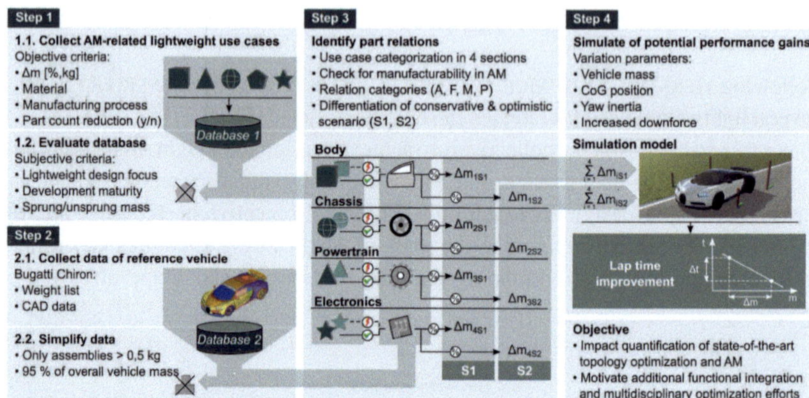

Fig. 5.2 Approach overview of potential assessment of AM-based lightweight design including a part-relation-analysis [TFV21]

Table 5.1 Objective and subjective attributes collected and assigned for use case data collection [TFV21]

Objective (quantitative)	*Subjective (qualitative)*
Application	Lightweight focus
Manufacturing process	Development maturity
Material	
Part count reduction	
Weight savings	

Detailed data of the respective vehicle in focus is also crucial for this analysis. The weight list and the CAD data compile the necessary vehicle related content for database 2 in Fig. 5.2. The amount of data is then simplified by excluding small assemblies ($< 0{,}5\,\mathrm{kg}$) to reduce overwhelming part count, thereby still considering 95 % of the vehicle's total mass.

Step 3: Identification of Part Relations

Relying on database 1 and 2 as input, in step 3 part relations between use cases and vehicle parts are identified. The technical vehicle sections relevant to this analysis are the body, chassis, drive train, and electronics. Database 1 and 2 were structured according to those technical sections (Fig. 5.2). Subsequently, four relation categories were defined to identify part relations between database 1 and 2 (Fig. 5.3).

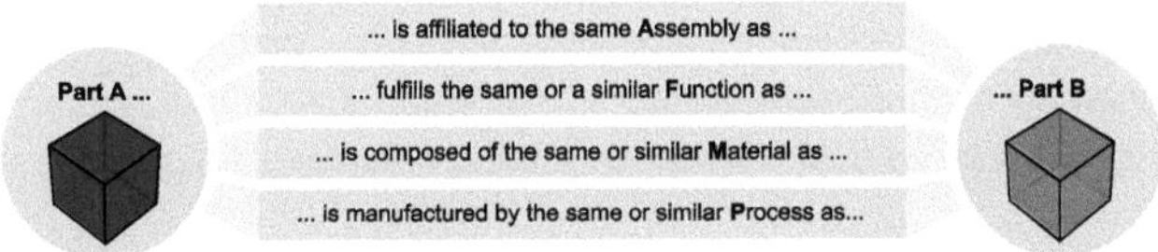

Fig. 5.3 Categories for the identification of part relations (A F M P) [TFV21]

Two scenarios were considered to account for inaccuracies during the identification of part relations and transfer of potential weight savings. The first scenario (S1) represents a conservative and the second scenario (S2) an optimistic perspective. Additionally, identified part relations are subjectively checked for plausibility in terms of the manufacturability of the part's geometry for both scenarios.

In conservative scenario 1, the highest weight savings are considered in case of multiple identified part relations. However, use cases gathered promise weight savings of 10 % to 75 %. Based on practical experience, high percentage weight savings can indicate a low development maturity or a low lightweight design focus of the original part of the respective use case. Therefore, the overall weight-savings for identified part relations in scenario one are capped at a maximum of 30 %. Based on [FFK19] a build envelope limitation of 1 m^3 is assumed for the respective AM process. Furthermore, the assessment of the lightweight focus of the AM use case and the corresponding vehicle component to which the weight-saving potential is transferred is taken into account in this scenario. For components of the target vehicle with a high conventional lightweight focus, only half of the additional use-case-related weight-savings were considered.

Scenario 2 also took the percentage weight savings of the part with the highest amount of part relations. Nonetheless, this scenario considers no maximum weight-saving threshold and no build chamber space limitation.

Step 4: Vehicle Track Performance Simulation Approach

Based on the identified part relations between database 1 and database 2 in step 3, the percentage weight-saving potentials are transferred to the reference vehicle (step 4 in Fig. 5.2). These weight-saving potentials add up corresponding to their technical sections Δm_i, resulting in

$$\Delta m = \sum_{i=1}^{4} \Delta m_i \tag{5.1}$$

as an input variable for simulation purposes. The lap time gains for various scenarios are determined and evaluated employing a vehicle dynamics simulation with the total weight savings and an analysis of further vehicle parameters, e.g., center of gravity (CoG) shift.

A validated model of the reference vehicle is necessary to simulate lap time alterations. For this purpose, the AVL VSM simulation environment and a validated vehicle model developed by Tschorn et al. (2020) in a previous study is chosen [TWV20]. The simulation technique is a driver model-based approach, for which the vehicle trajectory is determined in advance according to Gundlach and Konigorski (2019) [GK19]. The approximate 20.5 km long *Nordschleife* of the Nürburgring was chosen due to its well-known characteristics and comparability.

These weight saving potentials also impact other vehicle characteristics, such as the CoG position d_{CoG} and the yaw inertia I_z. To predict the potential shift of the CoG, an estimation based on an equilibrium of forces regarding the axle load distribution is elaborated. A basic calculation as shown in Fig. 5.4 is necessary to put the impact of weight reduction and its distance to the CoG on the vehicle yaw inertia into perspective. According to the estimation made with the approach above, the yaw inertia of the reference vehicle is adjusted.

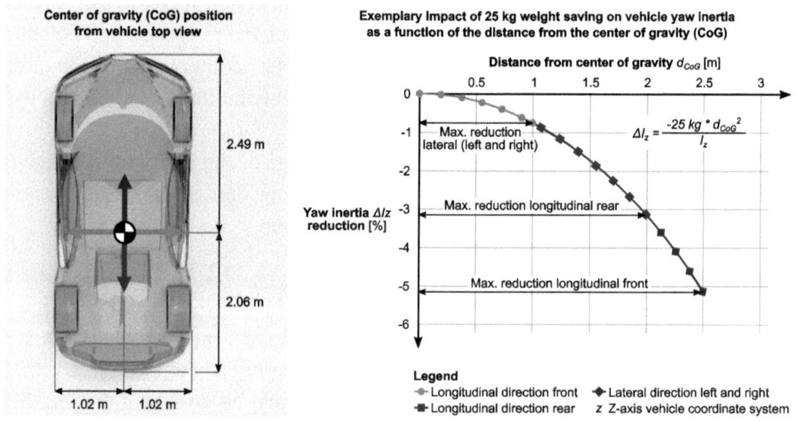

Fig. 5.4 Exemplary vehicle yaw inertia sensitivities for the reference vehicle [TFV21]

In addition to the considered parameter dependencies, weight-savings further improve vehicle performance. Here, the influence of adapted aerodynamics is also considered. For this purpose, the bearable downforce at top speed is increased, assuming the load capacity of the chassis of the initial reference vehicle remains the same. The necessary resulting alteration of the downforce coefficient Δc_D for the simulation is determined with equation (5.2).

$$\Delta c_D = \frac{2 \cdot \Delta m \cdot g}{\rho \cdot A \cdot v_{max}^2} \tag{5.2}$$

The additional downforce is divided according to the prevailing aerodynamic balance and assigned to the front $\Delta c_{D,F}$ and the rear axle $\Delta c_{D,R}$. In conclusion, weight reduction maintaining maximum load capacity of the chassis enables increased aerodynamic downforce of the vehicle while increasing drag too.

5.1.2   Findings

The main results generated in this analysis are derived from identifying part relations, the up-scaling of weight savings to a holistic vehicle level, and the track performance simulation, which are described in more detail in the following.

Identified Part Relations

This research gathered and analyzed 45 exemplary automotive applications in which weight savings were achieved due to topology optimization and AM. Those applications are mostly known to the public and communicated by OEMs or engineering service providers. Due to insufficient data, 16 use cases could not be considered. The 29 remaining examples focus on metal-only applications, consisting of either steel, aluminum, or titanium. The weight savings communicated in those exemplary applications differ between 10 % and 75 %. An exemplary selection of those use cases for the main technical sections of the vehicle is shown in Table 5.2.

As shown in Fig. 5.3, four different part-relation categories were identified, which are potential indicators for similarities between parts of different vehicles or assemblies. Based on these categories, part-relations throughout the reference vehicle were determined, a few exemplarily shown in Table 5.3. With this procedure, weight savings were firstly estimated and added up for each technical section and, after that, for the whole reference vehicle.

Leveraging of Weight Savings to the complete Vehicle Level

Table 5.3 shows one exemplary identified part relation between the use case and part/assembly of the reference vehicle for each of the four considered *technical sections*. The difference between *total mass* of the part/assembly of the reference

Table 5.2 Exemplary overview of automotive topology optimized AM applications and gathered data [TFV21]

Technical section	**Body System**	**Chassis**	**Drive train**
Use case	Space Frame Node [Eib16]	Brake Caliper [Bec18]	E-Drive Housing [Hof19]
Weight saving	25 %	41 %	32 %
Material	AlSi10Mg	Ti6Al4V	AlSi10Mg
Development maturity	Functional Prototype	Functional Prototype	Functional Prototype

Table 5.3 Examples of identified part relations and their projected weight savings for each technical section of the reference vehicle based on [TFV21]. (CW – considered weight, PRC – part relation category, S1 – conservative scenario, S2 – optimistic scenario)

Technical section Part/Assembly of reference vehicle	Total mass [kg]	CW [kg]	Identified part relation to use case	PRC	Weight reduction [%]	Weight savings [kg] S1	S2
Body						**11.80**	**22.68**
Frame section (front)	35.52	11.84	Frame-node [Eib16]	A F M P	25 %	2.96	8.88
Chassis						**18.42**	**21.06**
Wheel-drive (front)	13.77	10.31	Wheel-carrier	A F M P	15 %	1.55	1.55
Drive Train						**58.30**	**72.43**
Final drive (rear)	38.00	11.16	E-Drive housing [Hof19]	A F M P	32 %	3.35	3.53
Electrics						**1.40**	**2.31**
Various brackets	1.84	1.84	Bonnet hinge	F M P	23 %	0.42	0.42
Overall sum of weight savings						**89.92**	**118.50**

vehicle and the eventually *considered mass* describes the relation of parts that have been rated as *plausible* AM applications. Besides the *weight savings* for each *part/assembly*, that are derived from the *weight reduction* rates, the sum of *weight savings* for each *technical section* and the *overall sum* are also displayed in the two scenario-specific right-hand columns of the table. Overall, the projected weight savings based on the identified part relations sum up to around 89 kg in a conservative

(*S1*) and 118 kg (*S2*) in the optimistic scenario for the entire vehicle. According to this analysis, the total weight-saving potential adds up to approximately 4 to 6 % of the total vehicle weight.

The total result of the part relation analysis was transferred to the CAD model of the reference vehicle as illustrated in Fig. 5.5. Here, the location of the primary assemblies with weight-saving potential is visualized in the vehicle's design space. The technical sections of the vehicle are illustrated in different colors. The vehicle's front and rear segments, the drive train, and the chassis can be identified as areas with significant weight-saving potential. Note that this emphasis might only apply to vehicle concepts based on a fiber composite monocoque design.

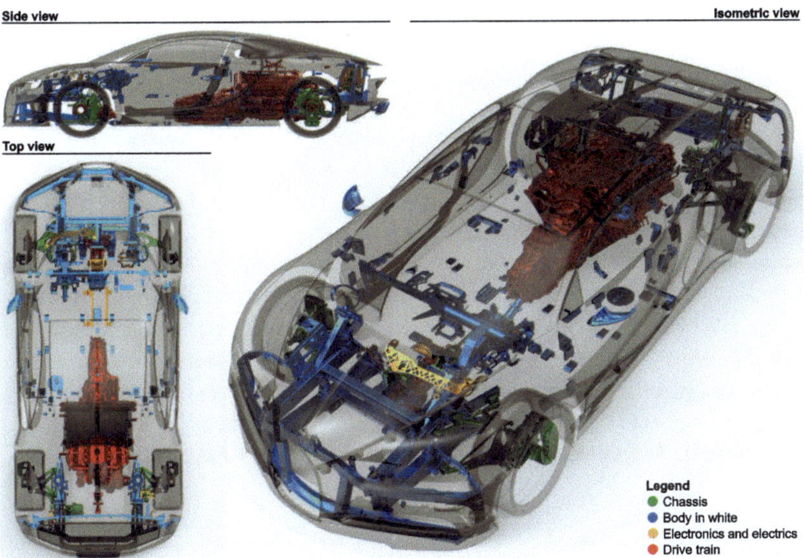

Fig. 5.5 Overview of identified vehicle parts potentially suited for AM-based lightweight design according to the main technical vehicle sections of a Bugatti Chiron [TFV21]

Track performance simulation

The lap time simulation results are displayed in Fig. 5.6. The weight savings (WS) and weight savings with adjusted aerodynamic downforce (WSA) have been simulated for scenarios 1 and 2. The overall vehicle weight has been gradually and iteratively reduced to obtain the general influence of weight reduction on the track

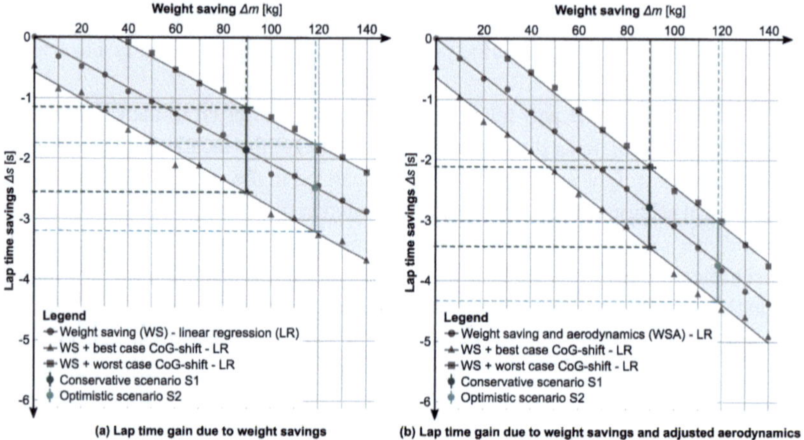

Fig. 5.6 Simulated lap time improvements due to AM-related lightweight design considering a conservative (S1) and optimistic (S2) scenario for reduced vehicle mass only (a) and reduced vehicle mass with equivalently increased aerodynamic down force (b) based on [TFV21]

performance of the reference vehicle (black round points in Fig. 5.6). Based on these points, a linear regression (LR) has been derived (black line). The impact of potential changes in the center of gravity (CoG) position in the longitudinal and vertical direction was also simulated. The resulting influence on the lap time savings is displayed by the square (worst case CoG-shift) and triangular points (best case CoG-shift) and their respective linear regression lines. For the adjusted vehicle parameters, based on the identified weight-saving potentials, two value pairs for lap time improvements were determined for each scenario. These are marked as two intervals (dark and light blue markings in Fig. 5.6) and are listed in Table 5.4.

Table 5.4 Potential lap time savings for scenario 1 and 2 considering weight savings only (WS) and weight savings with equivalent additional aerodynamic downforce (WSA) [TFV21]

	Conservative scenario (S1)	Optimistic scenario (S2)
WS	1.15 – 2.56 s	1.76 – 3.2 s
WSA	2.12 – 3.44 s	3.02 – 4.34 s

5.1.3 Summary and Discussion

The aim of the leveraging approach of this section is to assess the potential impact of AM lightweight design on a holistic vehicle level relying on resilient use cases. The vehicle lap time performance was chosen as the objective criterion and as a customer-relevant value proposition for this analysis. This analysis identifies the saving potentials for vehicle weight and thus the resulting lap time improvements on a holistic vehicle level with a focus on topology optimized AM manufactured metallic components. The identified potentials significantly impact vehicle weight and performance for a reference car with already emphasized lightweight properties. Therefore, the approach mentioned above turned out to be a suitable method to investigate the potential of a technical innovation on a vehicle in its entirety, particularly for weight savings and their impact on the vehicle's performance. Furthermore, this analysis enabled the identification of novel applications for topology-optimized additive manufactured parts in the reference vehicle, emphasizing that design freedom enabled by AM should be considered in track-performance-oriented automotive engineering.

Even though topology optimization methods are often associated with AM, weight savings might be achieved by combining topology optimization and conventional manufacturing processes. Nevertheless, considering the reference vehicle is a sports car, it has already been subject to a high lightweight focus during development and benefited from conventional lightweight methods. Noticeably, at least 50 % of the considered use cases are based on AlSi10Mg as material for structurally optimized vehicle parts. Even though this analysis sets no focus on material-based lightweight design strategies, further lightweight potential is conceivable considering alternative AM Materials with superior material properties such as e.g. Scalmalloy® [Awd+18]. Regarding the high costs of those alloys, they are primarily applied in the aerospace and healthcare industries. Further research is conducted to develop alloys suitable for automotive industry requirements, e.g., project *CustoMat 3D* [TEB20].

Moreover, the design freedom enabled by AM exceeds the capabilities to realize complex part shapes derived from topology optimization only, as many use cases nowadays mediate. Further, this analysis did not consider AM-related design potentials toward functional, hierarchical, and material complexity (Table 2.14). Additional lightweight design strategies (Fig. 5.1) and novel design methods such as functional integration and multi-physics design need to be considered to do so. To optimize the design of parts on a holistic vehicle level so that AM design potentials are fully leveraged, the right engineering tools and methods are essential. Those means are still to be developed to enable engineers to create innovative, interdisciplinary, and highly optimized designs that combine AM design potentials purposefully. Ten AM use cases considered in this analysis also achieved a part count reduction comparing conventional and AM designs. Nevertheless, additional weight-saving potentials arise if non-metallic parts are taken into consideration.

In order to reduce the complexity of this analysis, yet still be able to consider the entire vehicle, only assemblies with a mass greater than 0,5 kg were taken into account. The assemblies considered this way account for 95 % of the overall vehicle weight while ignoring numerous yet tiny parts. Even though a high lightweight potential was identified, the lightweight design focus of the gathered use cases could only be rated subjectively, which leaves room for error in the projection of weight reduction rates to the reference vehicle. Further, the development maturity of most of the use cases was classified as *prototype* or *functional prototype*. Hence, the use case weight savings might be lower when achieving the development maturity of a series part.

The aim of the chosen parameter adjustment approach for the lap time simulation is based on a comprehensive consideration of the complete vehicle. An additional outcome of this analysis is that the exact determination of the CoG shift is not possible when applying this technical up-scaling approach. Regarding the parameter dependencies between weight savings and CoG position, even with high-quality CAD data, due to the amount of considered assemblies without the final part geometry and thus the exact information about weight saving position, only an approximated estimation can be provided (Fig. 5.4). Further, the inconsistency of the simulation results (Fig. 5.6) occurs due to the chosen simulation technique of a driver model-based approach. The linear regression was performed concerning the linear behavior of the mass sensitivity and was considered a reasonable approach to display the lap time gains for the reference vehicle.

In conclusion, a holistic lightweight optimization taking advantage of AM design potentials can impact overall vehicle performance independent of reasonable manufacturing costs. The approach mentioned above successfully explored the potential impact of a technical principle on a holistic product, as defined in this chapter as technical up-scaling. In particular, the conducted identification of part relations was a helpful principle in identifying appropriate applications with weight-saving potential on a holistic vehicle level, enabling the design engineer to prioritize those parts for further improvement. With respect to the identified weight savings, the simulation results show possible lap time reductions up to four seconds on the *Nürburgring Nordschleife* if aerodynamic downforce adjustments are taken into account. These lap time savings are significant but should be assessed in relation to the race track length, which exceeds 20 km. To utilize the full potential of AM-design freedom, design solutions focusing on functional integration and multi-disciplinary optimization are indispensable and, therefore, necessary. Functional integration has the potential to save even more weight, e.g., through part count reduction. It can also increase vehicle performance regarding interdisciplinary functions such as heat management, increasing vehicle durability and consistency. Further, collateral impact on other vehicle properties is challenging to determine. However, geometric changes due to part count reduction and additional weight savings due to the resulting causality chain (reversed *weight spiral*) are possible. As an additional example, the handling properties are also positively affected by reducing vehicle mass and inertia as one of many customer-relevant value propositions.

Lastly, the main findings of this section are quantified weight-saving potentials for a complete vehicle. This analysis considered only geometry-based lightweight design principles combining topology optimization and AM-related design freedom for metallic parts. The weight-saving rates, derived from a gathered database of AM lightweight use cases, were transferred to a reference vehicle based on identified part relations elaborated with the introduced technical up-scaling approach. Significant weight reduction potentials for the reference vehicle resulted. The impact of the derived weight savings on the track performance of the reference vehicle is determined through lap time improvements employing a vehicle dynamics simulation. The results achievable with the introduced technical up-scaling approach can be considered an educated and estimated quantification of the disruptive potential of metal-based AM technologies on vehicle track performance. Thus, with these limitations in mind, this section elaborated on the holistic impact of the first key application area of AM for vehicle conception *increase vehicle performance* with a focus on lightweight design only (see Fig. 3.7).

5.2   Approaches to AM-based Vehicle Body Design

Regarding the overview of design strategies towards conventional vehicle body structure summarized in Sect. 2.2.2, it becomes clear that the five main established approaches are suitable for specific vehicle types, e.g., ladder frame design for heavy-duty commercial vehicles (compare Table 2.9). Each design principle is based on a different set of conventional manufacturing processes, e.g., welded extrusion profiles, sheet metal forming, casting, or lamination of fiber composites. The state-of-the-art AM vehicle concepts discussed in Sect. 3.1, the AM application principles generated in Sect. 3.2.3, the individualization-based use cases of Sect. 4.1, and the functionally integrated vehicle segment of Sect. 4.2.2, stress the necessity for unconventional design approaches to leverage AM design potentials fully. Further, the findings of Sect. 5.1 show that part-level focused topology optimizations take advantage of AM-related design freedom for the holistic vehicle only to some extent. In the context of this thesis and considering the aspects mentioned above, the question arises if additional general approaches to structural body design might be necessary to holistically exploit DfAM potentials for VC. This section elaborates on the related work (Sect. 5.2.1), on the systematic generation of novel approaches to vehicle body design (Sect. 5.2.2), and on their evaluation (Sect. 5.2.3).

5.2.1   Related Work

As shown in Sect. 3.1, several approaches to AM-based vehicle design can be found in the literature and other means of publication. The gathered concepts address a wide range of vehicle concept characteristics based on DfAM potentials, starting with *show cars*, *low requirement vehicles*, and *individualization focused concepts* up to *high performance*, and *generative concepts*. However, those concepts known to the state-of-the-art are primarily based on established approaches to vehicle body design, which again are based on the design freedom and restrictions of conventional manufacturing techniques (compare Fig. 3.2). A systematic analysis of additional design approaches to vehicle body design taking advantage of AM-related design freedom and restrictions could not be found in the literature.

Nevertheless, additionally to the topology optimization-based approach from the *3i-PRINT* project thoroughly discussed in Sect. 4.2.1 [CSI18], contemporary research addresses several aspects of AM-based vehicle body design. In the Bio-LAS project, e.g., JAESCHKE ET AL. (2020), the feasibility of bionic LPBF-based lightweight design for automotive applications with a focus on material, design and hybrid manufacturing has been analyzed. Here, parameter development of a cost-effective steel alloy, the definition of design-to-cost design guidelines, and joining techniques between conventional and AM parts were in focus [Jae+20]. Further, LI AND KOHLWES (2021) elaborated on increasing process productivity of steel-based LPBF for automotive body applications [LK21]. To increase the LPBF process's output and reduce the respective costs, *Seurat Technologies* is developing large-area pulsed LPBF [Roe+21]. Furthermore, an AM-specific aluminum alloy has been developed in the project *CustoMat3D* to fulfill automotive crash requirements [Lut23]. Also, additively manufactured structural automotive parts based on hierarchical complexity utilizing lattice structures for crash purposes have been patented, e.g., by General Motors Company [MPW18]. However, the applications considered in those research efforts mentioned above are limited to small build volumes and do not address the vehicle in its entirety (e.g., rear axle mounting and steering column bracket [Jae+20, pp. 97, 99], or front suspension strut tower [Lut23, p. 130]).

Several AM vehicle concepts can be assigned to conventional body design techniques when analyzing the AM vehicle concepts gathered in Sect. 3.1 and classified in a correlation matrix (compare Fig. 3.2). However, the concepts mentioned in the following are not as easily classifiable as only one conventional technique. For example, the show car *Light Cocoon* (EDAG) is based on a space frame design consisting of profiles and additive manufactured metal nodes. Nonetheless, the outer bionically inspired polymer body skeleton is covered with cloth material, closing open skeleton voids yet enabling backlit functionalities (compare Table 3.2). The *Strati* (Local Motors) is based on a one-piece short fiber reinforced polymer body printed in one process step unifying monocoque and body panels showing similarities to a unibody shell design (compare Table 3.1). The divergent concepts *Blade II and III* show characteristics of a space frame and a monocoque-focused approach (compare Table 3.5). In Addition to those concepts, the white spots shown in the correlation matrix in Fig. 3.2 can also indicate technical limitations to AM-based vehicle concepts, such as a polymer-based space frame. Otherwise, those white spots can also foster the "ideation" of novel approaches.

5.2.2   Concept Generation

Considering the aim to leverage DfAM potentials to the complete vehicle level, this subsection elaborates on differing approaches to AM-based structural vehicle body design. For this purpose and to question conventional approaches, this section does not consider current AM technology limitations based on build size restrictions or batch-related production cost structures. Further, no particular vehicle classes are excluded for the concept generation phase, as different resulting approaches might be suitable to various types of vehicles. The approach chosen for this analysis is based on the following three significant aspects:

- **Functional Requirements:** Prioritized set of functional requirements relevant to vehicle body structures
- **Manufacturability:** Combination of manufacturing possibilities considering conventional and AM technologies
- **Flexibility:** Considering individualization potential due to flexibility of toolless manufacturing techniques

The functional requirements of a vehicle body are prioritized and reduced to the most fundamental requirements for this concept generation stage. The resulting prioritized requirements are listed in Table 5.5.

Table 5.5 Prioritized requirement sets for AM-based vehicle body concept generation

Vehicle Body Requirements
Cluster
   Set
      Subset
	Priority	Vehicle Body Requirements
		Cluster
		   Set
		      Subset
			Priority

Rigidity and strength:		Interfaces and mounting points:	
Crash safety:	•	Drive train	•
Passenger compartment	•	Chassis	•
Crumple zone	•	Interior	•
Pedestrian protection	•	Windows	•
Torsional stiffness:		Integration level of hang on parts	•
Acoustic and vibration comfort		Design, package and shape:	
Driving dynamics		Driver Visibility	•
Fatigue strength	•	Ergonomics	
Resistance to weathering and corrosion		Anthropometry	
Ease-or-repair		Styling	
Recyclability		Aerodynamics	
Surface finish:		Cost effectiveness	
Smoothness		Lightweight design	
Painting and coating suitability			
Styling features			

Based on those functional aspects shown in Table 5.5, four ideation workshops were carried out to creatively gather potential concept ideas among a small group of four experts. The experts have profound expertise in mechanical engineering, VC, and DfAM. Besides the ideation of concepts, their categorization and assessment are in focus. The resulting AM-based design approaches to structural body design are summarized in Fig. 5.7 in a generalized manner. The generated principles are ordered according to technological feasibility regarding implementation difficulty and respective period for realization. The first approaches start modestly, in which conventional elements remain the major load-bearing components of the structure (*principle A–E*). However, with increasing AM penetration level, space frame nodes and/or struts are suggested to be additively manufactured. Nonetheless, based on the high load-related requirements valid for automotive vehicles, most approaches consider a combination of metallic structures for load-bearing body components and polymer shell elements for fairings and panels (*principles A–F*). Only an approach based on a continuous fiber-reinforced polymer structure does not rely on metallic elements (*principle G*). Even though the listed approaches in Fig. 5.7 might not be exhaustive, most already demonstrated AM principles in the literature are covered (*principles A–D, F*), whereas additional principles not yet demonstrated in the literature where gathered (*principles E, G, H*) including multi-material-based AM technologies (*principles G–H*).

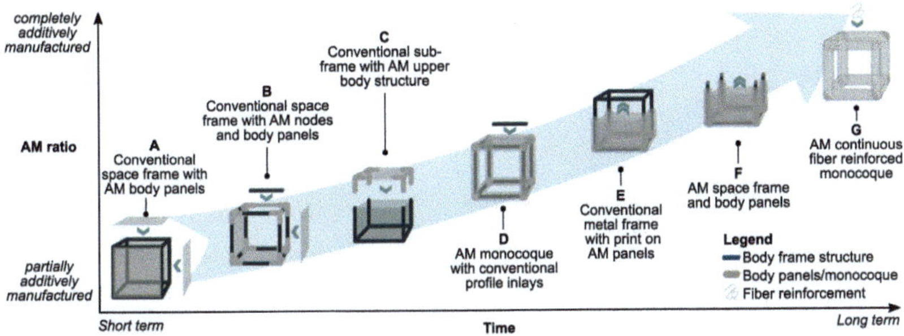

Fig. 5.7 Overview of AM-inspired body design approaches in order to their time-related feasibility

Those principle approaches to AM-based structure design are subsequently applied to a vehicle body to illustrate potentially resulting body architectures exemplarily. Table 5.6 summarizes those concepts to clarify potential embodiments of the principles illustrated in Fig. 5.7. A few of these are described in further detail below. The letter in the concept name refers to the respective design principle summarized in Fig. 5.7 on which the concept is based. Variations of the same basic principles have been found if a number follows the letter. For visualization purposes, the AM-based body concepts illustrated in Table 5.6 resemble conventional vehicle topologies (Table 2.8). This circumstance should not indicate that differing vehicle topologies and overall shapes are not possible or even more suitable. Further, the illustrated concepts do not focus on AM-related functional integration potential already elaborated in Sect. 4.2. The main objective of the suggested concepts is to achieve structural integrity according to automotive functional requirements (Table 5.5).

However, depending on the VC characteristics (compare Fig. 2.21), the gathered approaches might be suitable for opposing vehicle classes. For example, on the one hand *concept A* finds use for low requirement yet highly individualized vehicle concepts (e.g., *Micro Commuter* in Table 3.3). On the other hand, *concept B* shows similarities to the layout of high-performance vehicles with a focus on lightweight design (e.g., *Blade III* in Table 3.5). Accordingly, different requirement levels apply to the AM share of the respective generated concepts. With increasing mechanical requirements on AM parts, the necessity for either reinforced polymer

Table 5.6 Overview of exemplary AM-inspired vehicle body architectures based on design principles of Fig. 5.7. (For comprehensive list compare ESM App. A.7)

A. Conventional space frame with AM body panels

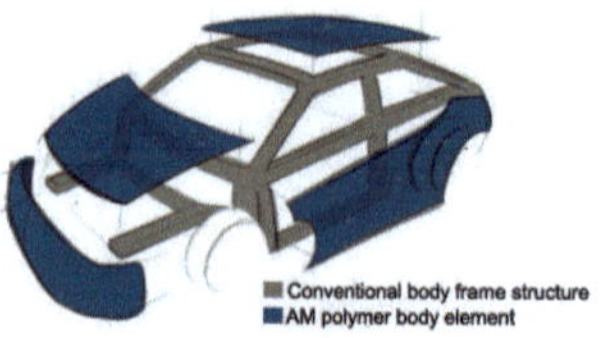

B. Conventional space frame with AM nodes and body panels

C1. Conventional sub-frame with AM upper body structure

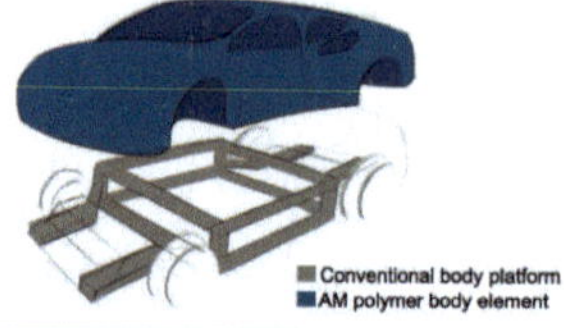

C3. Conventional sub-frame with commercial AM upper body structure

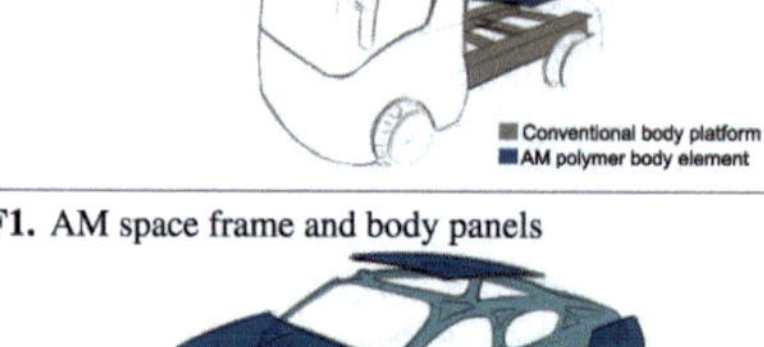

D. AM monocoque with conventional profile inlays

F1. AM space frame and body panels

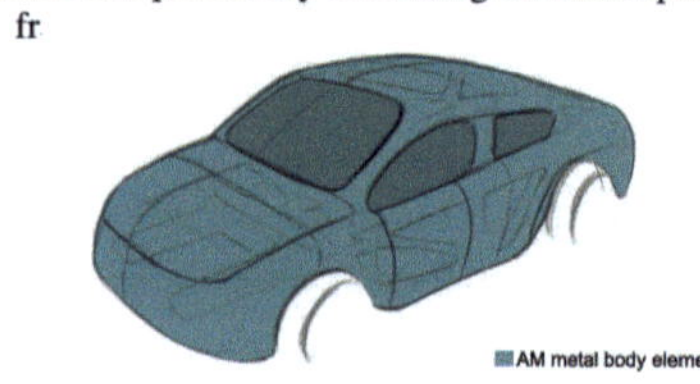

F2. AM space frame and body inserts

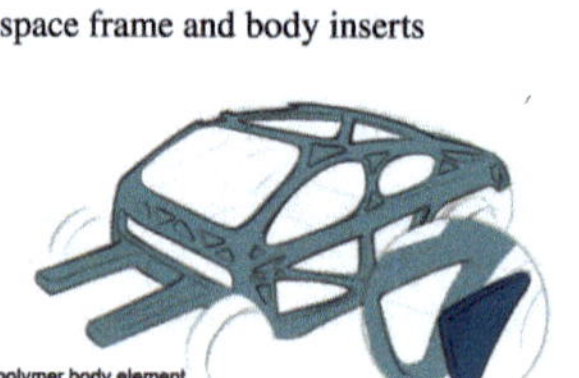

F3. AM one-piece body combining shell and space fr

F4. AM one-piece body combining shell and sandwich

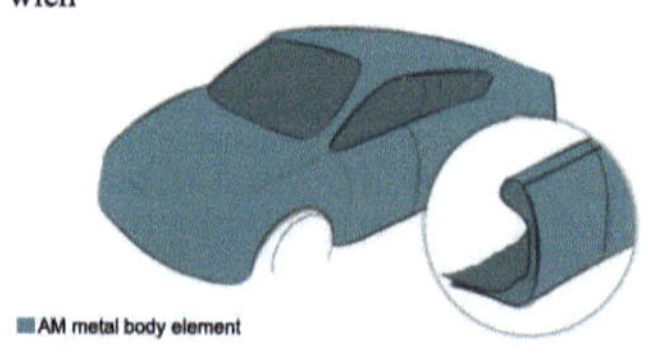

G. AM continuous fiber reinforced monocoque

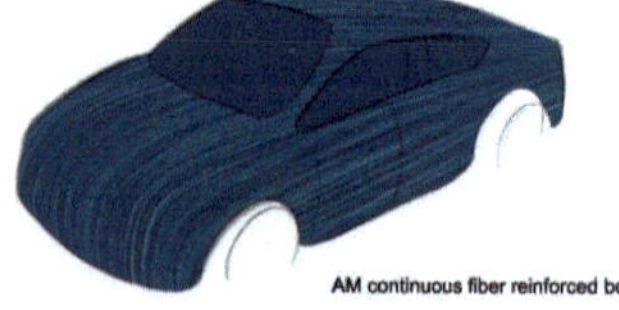

or metallic materials increases. Subsequently, manufacturing costs are expected to increase respectively, limiting the suitability of approaches such as *concepts F1–G* to premium high-performance vehicle applications.

5.2.3   Concept Evaluation

The AM-related vehicle body concepts generated above are further evaluated in this subsection. This way, it is possible to assess each approach's advantages, disadvantages, and general characteristics. For this purpose, each concept has been rated by the workshop participants according to four primary assessment criteria. Each of which is divided into five fulfillment increments between *low* and *high* (0, 25, 50, 75, and 100 %). The assessment criteria are described in more detail as follows:

- **AM ratio:** This criterion expresses the penetration level of AM processes for the respective vehicle body. A full rating accounts for a high up to 100 % share of AM parts necessary to realize the approach.
- **Feasibility:** A concept's feasibility level considers aspects such as the availability of necessary means of production, an approach already demonstrated in state-of-the-art, or general technical challenges to be considered. A maximum rating accounts for the high feasibility of the approach when considering the technological maturity of the necessary manufacturing processes. In this case, no technological leap is necessary for further concept realization. In contrast, a low rating suggests that without major technological leaps in the respective AM process, the approach might be only feasible with unreasonable additional effort in post-processing and, therefore, high manufacturing costs.
- **Requirements:** This rating considers the difficulty level of requirements applying to the AM components of the concept. A high rating accounts for crucial crash-safety relevant components subject to high static and dynamic loads. Low-rated components do not contribute significantly to the vehicle's structural integrity. Please note that, e.g., the surface finish has not been prioritized in Table 5.5. Therefore, high requirements for the look and feel of the vehicle are not addressed in this rating.
- **Individualization:** To a certain extent, this criterion also depends on the first criterion *AM ratio*. As with increasing AM penetration level, less tool-based manufacturing technologies are necessary, which usually decrease manufacturing flexibility and thus the individualization potential of the approach. However, depending on the design strategy, some AM-focused approaches are more affected by individualization-related changes in geometry than others, resulting

in more or less engineering complications. A high rating here expresses a high individualization potential of an approach. On the contrary, a low rating indicates strong dependencies on conventional tool-based manufacturing technologies.

- **Innovation level:** This criterion indicates whether state-of-the-art use cases have already demonstrated the approach. A high rating indicates a novel, uncommon approach to AM-based vehicle body design.

As the description of each concept would exceed the scope of this thesis, three characteristic approaches that differ significantly from each other are discussed in more detail. Please compare App. A.7 for the detailed concept descriptions and evaluation results.

Concept B illustrated in Table 5.7 consists of a metallic space frame structure compiled from conventionally manufactured structural extrusion profiles and additively manufactured nodes. Space frame nodes often rely on complex geometries to connect structural beams from various angles. Here, AM-related design freedom might be helpful to, e.g., realize complex lightweight geometries or facilitate the generation of derivatives. Besides the metallic space frame, fairings and panels are suggested to be manufactured by polymer-based AM technologies. This way, costs might be reduced as these components are mostly not subject to extensive loads. Combining toolless manufactured nodes and fairings with tool-based yet lengthwise variable extrusion profiles bares some individualization potential in the vehicle architecture. Depending on the chosen AM processes, this concept outlines a compromise between manufacturing costs and individualization potential. Thus, costly manufacturing processes such as LPBF are limited to only geometrically complex

Table 5.7 Description and assessment of AM-inspired body design approach B

Concept B Conventional space frame with AM nodes and body panels		*low*	*high*
AM ratio		▮▮	
Feasibility		▮▮▮	
Requirements		▮▮▮	
Individualization		▮▮	
Innovation level		▮▮	

Advantages and disadvantages:
✓ Compromise between costs and individualization
✗ Thermal expansion differences
✗ High surface finishing effort

yet small parts. For this aspect, this approach is considered to be rather feasible, as it already has been partially shown in state-of-the-art approaches (e.g., *Light Cocoon* in Table 3.2 or *Blade III* in Table 3.5). However, engineering challenges such as thermal expansion differences between materials and surface roughness of polymer manufacturing processes such as SLS and FDM need further attention (compare Fig. 2.30 for comprehensive AM process overview).

In Table 5.8 *concept F2*, a variation of *concept F*, is illustrated and can be seen as an evolution of *concept B* in regard to the integration level. Here, the basic body structure is also based on a space frame approach, yet the entire metallic frame structure is manufactured additively. Thus, strut-like elements are integrated into the composition of AM nodes and not conventionally manufactured.

Table 5.8 Description and assessment of AM-inspired body design approach F2

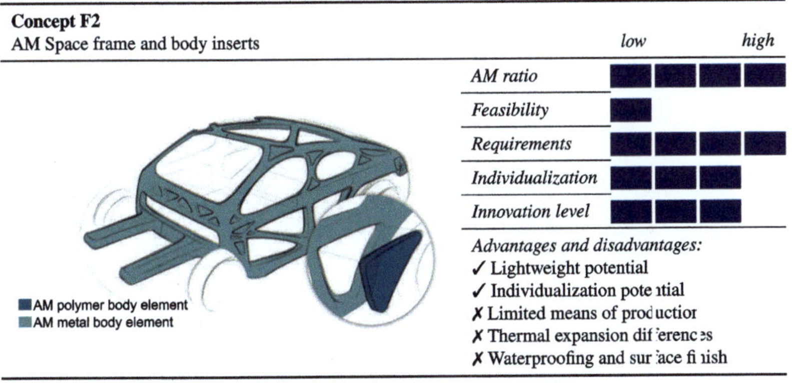

The body structure in this concept is not meant to be covered by body panels and fairings. Instead, the structure's openings are closed with bespoke polymer AM inserts similar to a timbered house design, thereby integrating fairing components into the space frame. This composition results in a 100 % AM ratio as no conventional manufacturing processes are considered. The feasibility of this approach is rated rather low as means of production for such large AM parts are limited or necessary joining techniques unreasonable as the AM structure must comply with high safety requirements. However, this technique bares advanced lightweight and individualization potential as highly optimized shapes throughout the vehicle structure are possible without tool-restricted manufacturing processes. Further, this approach has not yet been observed in state-of-the-art vehicle concepts and is therefore considered innovative.

Concept F3 illustrated in Table 5.9 represents a fully integrated variant of *concept F1*. Here, a one-piece metallic vehicle body that combines space frame and shell design is targeted. The vehicle body's outer shell is not separated from the structural frame. Shell-like elements are rather stiffened with integrated strut-like reinforcements. The entire integrated body is supposed to be additively manufactured. The combination of frame and shell design promises high structural body stiffness and low body mass as panels can be considered load-bearing elements, reducing stress on and, thus, the bulk of strut-like body sections.

Table 5.9 Description and assessment of AM-inspired body design approach F3

Concept F3

AM one-piece body combining shell and space frame design

	low	high
AM ratio		(high)
Feasibility	(low)	
Requirements		(high)
Individualization		(high)
Innovation level		(high)

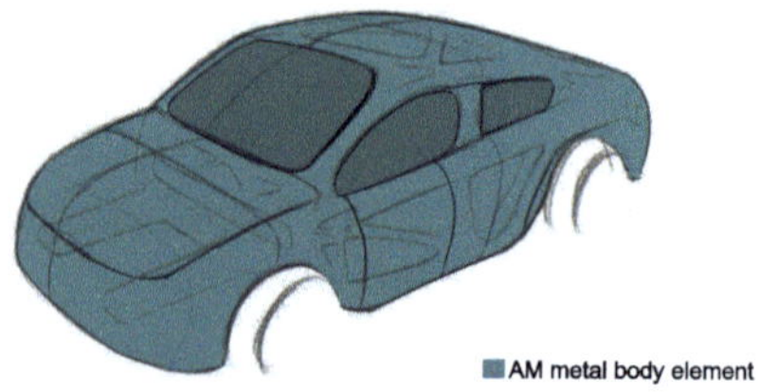

AM metal body element

Advantages and disadvantages:
✓ Lightweight potential
✓ Individualization potential
✗ Limited means of production
✗ Costs
✗ Ease-of-repair
✗ Surface finish

Nevertheless, the feasibility of this approach is rated low due to the limited means of production for such part dimensions and the challenge of manufacturing accurate, thin, and sheet-like parts with a high aspect ratio with most metal-based AM processes. This aspect occurs from thermal distortion issues during the part build-up (e.g., LPBF) or due to shrinkage while part sintering (MBJ). Further, this context impedes the fulfillment of high safety-related requirements. However, high lightweight and individualization potential is assigned to this approach even though challenges regarding cost-effectiveness, ease-of-repair, and surface finish need to be considered. Even though a similar polymer-based approach to monocoque design can be observed in the state-of-the-art, e.g., *Strati* by *Local Motors* in (Table 3.1), this metal-based approach is considered innovative as it might enable high-performance AM vehicle concepts instead.

The additional concepts of Table 5.6 not discussed in detail also imply different aspects of AM-enabled vehicle body design with specific advantages and disadvantages. Table 5.10 summarizes the evaluation results of all generated concepts.

Table 5.10 Overview of evaluation results AM-inspired body design approaches

Approach	Concept	AM ratio	Feasibility	Requirements	Individualization	Innovation level
A	A: Conv. space frame + AM panels	■□□□	■■■□	■■□□	■□□□	■□□□
B	B: Conv. space frame + AM nodes and panels	■■□□	■■■□	■■■□	■■□□	■■□□
C	C1: Conv. sub-frame + AM upper body	■■□□	■■□□	■□□□	■■□□	■■□□
C	C2: Conv. sub-frame with rollover protection + AM upper body	■■□□	■□□□	■■□□	■■□□	■■□□
C	C3: Conv. sub-frame + commercial AM upper body	■□□□	■□□□	■■□□	–	■□□□
D/E	D/E: AM monocoque + conv. extrusion profile inlays	■■■□	■□□□	■■□□	■■■□	■■□□
F	F1: AM Space frame + body panels	■■■■	■□□□	■■■■	■■■■	■■■□
F	F2: AM Space frame + body inserts	■■■■	■□□□	■■■■	■■■□	■■■□
F	F3: AM one-piece body in shell + space frame design	■■■■	■□□□	■■■■	■■■■	■■■■
F	F4: AM one-piece body in shell + sandwich design	■■■■	■□□□	■■■■	■■■■	■■■■
G	G: AM fiber reinforced monocoque	■■■■	■□□□	■■■■	■■■■	■■■■

Each vehicle design approach generated in this section might suit differing vehicle types, ranging from low requirement to commercial vehicles up to high-performance sports cars. Even though the evaluation summarized in Table 5.10 is subjective, certain central tendencies can be observed that also correlate to the main characteristics of AM. First, with increasing *AM ratio* in the vehicle concepts A to G, the dependence on tool-based manufacturing processes decreases, and therefore the individualization potential also increases. Second, also in correlation to the increasing AM ratio, the innovation level rises too, indicating approaches not yet shown in the state-of-the-art concepts. Third, the requirement level correlates reciprocally to the concept feasibility, indicating that nowadays it is challenging to efficiently meet high requirements with additively manufactured safety-relevant parts.

Further, regarding *concept C3*, it is important to note that an additively manufactured upper body would enable high individualization options for, e.g., commercial use cases. However, as this individualization potential only impacts the vehicle's upper body and not the vehicle concept itself, a high individualization rating here would not be comparable to ratings from other concepts and is therefore not considered in this evaluation step.

5.2.4 Summary and Conclusion

Different AM-based vehicle body design approaches have been generated and evaluated. First, state-of-the-art approaches and related research have been discussed, emphasizing the necessity for a systematical elaboration on this topic (Sect. 5.2.1). Next, a three-step approach to concept generation based on ideation workshops with AM and VC experts has been conducted (Sect. 5.2.2). Initially, the scope of fundamental functions of vehicle body engineering has been prioritized to reduce complexity for a high-level concept generation (Table 5.5). Seven principle approaches to AM-based vehicle body design have been gathered, including state-of-the-art and novel approaches (Fig. 5.7). Based on those seven principle approaches, eleven vehicle body concepts have been derived combining conventionally and additively manufactured vehicle components (App. A.7). Three material categories, metals, polymers, and fiber-reinforced materials, are considered. Lastly, the generated concepts were evaluated concerning the five aspects *AM ratio, feasibility, requirement level, individualization potential*, and *level of innovation* (Table 5.10).

The generated concepts illustrate creative and novel approaches to structural vehicle body design, challenging conventional automotive design methods and indicating additional AM-related potentials to VC that are not yet feasible with conventional manufacturing technologies. Furthermore, each approach might be suitable for a specific niche of vehicle classes, types, or intended purpose of use.

A quantified assessment of the major characteristics of each concept would be necessary to finally identify the most promising approaches to AM-based vehicle body design for a respective vehicle class or target customer group. However, the generated concepts are not tangible enough to quantify their properties at this stage. Further, a detailed analysis of each concept's identified engineering challenges is necessary for a quantified evaluation. The differing characteristics of the multitude of AM processes should be considered in the next step. Moreover, most generated concepts show a low feasibility rating even though the final AM processes for each concept have not been defined (Table 5.10). It becomes clear that in the opinion of AM and VC experts, AM technologies generally demand further development progress concerning quality, reproducibility, productivity, and build volume to increase the feasibility of safety-relevant vehicle components. Only then AM technologies can be considered for vehicle concepts with a high AM ratio.

Impact on Vehicle Development Methods and Tools

6

This chapter elaborates on the impact of design for additive manufacturing (DfAM) application approaches on vehicle development methods and tools. With the identification of the key application areas of AM on VC in Chap. 3, the analysis of specific automotive applications on part-level in Chap. 4, and the leveraging of AM characteristics to complete vehicle level in Chap. 5, the potential impact of AM-related design freedom on VC has been outlined.

The three significant findings from the elaborations on design methodology above are synthesized in this chapter. First, in Sect. 6.1, the aspect of product-benefit-oriented application of AM design potentials is summarized as a methodological approach to product design centered on conflicting objectives. Second, the restrictive aspect of engineering tools is summarized in Sect. 6.2 indicating necessary advancements in CAD/CAE to enable design engineers to leverage AM design potentials fully. Third, a multiphysics optimization (MPO) design approach is suggested as a general design strategy as opposed to the conventional hands-on CAD design process in Sect. 6.3.

6.1 Approach for Product-Benefit-Oriented Application of AM Potentials

As a result of its unique design and manufacturing freedoms, AM enables innovative product designs and business models. A customer-benefit-oriented use of AM potentials should be targeted to apply AM potentials purposefully and economically in the long term. This way, additional design freedom is considered in the early stages of the PDP, benefiting the product's overall performance. As shown in more detail in the following, this aspect motivates the field of research *Design for Additive Manufacturing* (DfAM). As summarized in Sect. 2.3, DfAM pursues the goal of

© The Author(s) 2026

183

D. Fuchs, *Exploration of the Influence of Additive Manufacturing Design Potentials on Vehicle Conception*, AutoUni – Schriftenreihe 179,
https://doi.org/10.1007/978-3-658-49677-7_6

clarifying AM-related freedom and restrictions in design and production, suggesting approaches, methods, and tools for knowledge provision and application. By developing suitable methods for the DfAM processes, customer-benefit-oriented use of AM design potentials can be achieved, thus promoting the creativity of design engineers and enabling efficient and effective use of those potentials [Kum+18; YPZ18; Ric+18].

Currently, design engineers familiar with conventional manufacturing processes (e.g., milling, casting, and molding) sometimes need to gain the necessary knowledge to develop products utilize AM potentials purposefully, considering them in the early stages of the product development process. One possible consequence is that applying individual AM design potentials, such as complex lattice structures, undercuts, or bionic shapes, does not always benefit product performance [Ric+18; Pra+18].

This section presents a product-benefit-oriented approach for applying AM potentials based on conflicting objectives in product requirements to counteract this issue. Thereby, available DfAM methods are taken into account. The derived approach is evaluated by means of academic and industrial expert workshops. With the assistance of the introduced approach, design engineers shall be equipped with a guideline for a product-benefit-oriented application of AM potentials. Additionally, helpful tools in the specific development phases of the product are pointed out. In Sect. 6.1.1 the approach is described in detail. Section 6.1.2 explains and discusses the conditions of the conducted industrial and academic expert workshops. In Sect. 6.1.3, the findings are summarized. Note that additional elaborations on this approach can be consulted in KUSCHMITZ, FUCHS, AND VIETOR (2023) [KFV23; Kus23, p. 69].

## 6.1.1	Approach Description

As a result of AM design possibilities, complex lattice structures, undercuts, or bionic-shaped geometries can be manufactured without significant additional effort during production [Kum18; GH16]. However, these potentials should only be purposefully integrated into product design. Otherwise, the AM implementation does not optimize the part's performance concerning its requirements. For example, solely additively manufacturing a brake caliper geometry initially designed for a casting process without a redesign most likely will not lead to an increase in part's performance nor decrease its manufacturing costs, this way not leveraging AM design potentials holistically [KFV23].

To enable a product-benefit-oriented implementation of AM potentials in the conception and design process, the research field DfAM provides the necessary information and tools for design engineers during product development. However, as emphasized in Sect. 3.3, the application of AM potentials might not have only beneficial effects on the properties of a product. Combining DfAM tools enables design engineers to develop a holistic understanding of the product-benefit-oriented application of AM potentials. As outlined in Sect. 2.2.4, a product, or in this case, a vehicle, represents the best compromise between many conflicting requirements and development objectives design engineers could achieve within their abilities and resources. This compromise includes the restrictions of available engineering tools and manufacturing technologies. With this in mind, it is assumed that additional design freedom can lead to improved trade-offs between conflicting development objectives for the respective vehicle concept leading to customer-relevant product benefits. Hence, the approach described in the following focuses available DfAM methods on improving trade-offs in opposing product properties. This approach should be applied in the early stages of the product development process.

Figure 6.1 illustrates an overview of the developed approach with its main execution phases, including exemplary DfAM methods to do so. A *market pull* or a *technology push* is assumed as potential initial triggers for applying AM potentials.

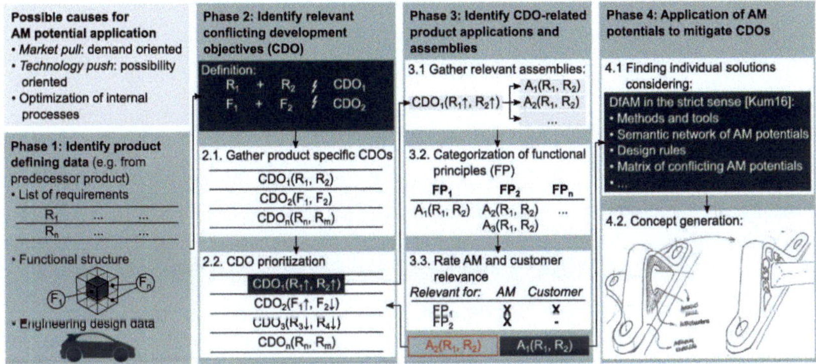

Fig. 6.1 Approach for a product-benefit-oriented application of AM potentials focused on conflicting development objectives (CDO) based on [KFV23]

Phase 1: Collection of product defining information

In the first phase, if available, the necessary information about the product, the assembly, or the part to be developed is gathered. The design task might be an original, adaptive, or variant design [Pah+07, p. 4]. The product specification, including the list of requirements or specifications, the functional product structure, and existing 3D CAD data, such as the usable design space, technical package models, and possible predecessor parts, are considered relevant information.

Phase 2: Determination of conflicting objectives due to opposed product requirements

In the second phase, after the relevant information has been gathered, all product requirements and derived functions are subsequently examined to identify possible conflicts arising from contradictions (see Fig. 6.1, phases 2.1 and 2.2). In this phase, the product requirements are assessed for trade-offs independently of existing component partitions, ensuring that conflicting requirements are identified regardless of conventional part geometry and the functional scope of predecessor designs. In this context, the definition of conflicting objectives by EILETZ (1999) is recommended (Sect. 2.2.4). This approach does not limit the solution space due to conventional manufacturing restrictions. In phase 2.3, the identification of relevant trade-offs and diverging requirements is concluded with the prioritization of identified trade-offs. The prioritization should be based on the relevance of the identified trade-off for the overall product performance from a customer perspective [KFV23].

Phase 3: Identification of related product assemblies, functions or parts

In the third phase, the identified conflicting objectives on the product level are assigned to affected components and assemblies (see Fig. 6.1 phase 3.1). Those related assemblies are categorized into basic functional principles necessary to realize product-specific features, e.g., kinetic energy absorption, fluid, heat transfer, and vibrational resistance. This way, the suitability of AM for a specific assembly can be assessed based on functional principles to be realized and not only on the embodiment of predecessor parts. Subsequently, in Fig. 6.1 phase 3.3, a classification of the suitability of AM design principles for optimizing identified components is assessed. The focus of this assessment is, in general, the compatibility of the component-specific functional principles with general AM characteristics based on DfAM knowledge. Assemblies with low suitability to AM potentials and restrictions (e.g., build volume, accuracy, or material properties) are not further considered. Otherwise, identified components are considered for the following conception phase. The analysis is intentionally carried out on component/assembly, not part-level. This way, opportunities for functional integration and part consolidation are not excluded in the early stage of the analysis [KFV23].

Phase 4: Application of AM potentials to optimize trade-offs

In the fourth phase, to utilize the AM optimization potential for the identified components and assemblies as holistically as possible, it is required to rethink and, if necessary, redesign the component of interest, including its assemblies and parts, with support of, e.g., the indicated DfAM tools. Those DfAM methods indicated in Fig. 6.1 are particularly suitable for the development of conceptual solutions. After completion of conceptual solutions, part design concepts are developed based on identified AM potentials and their relations among each other (see Fig. 6.1, phase 4.2). As a result, a design concept is derived, which addresses customer-relevant product requirement trade-offs suitable for a specific AM process and systematically considers AM potentials. The derived concept is suggested for further detailed development activities according to the respective PDP [KFV23].

The exemplary concept illustrated in Fig. 6.1 shows a conventional component mount on the left. This mount resembles a compromise between at least the three development targets, e.g., withstanding component loads, damping component-induced vibrations from the vehicle body, and minimizing weight. The conventional design is based on a two-component concept consisting of an elastomer and a rigid mount section to solve the conflicting load-bearing and vibration-damping objectives. In this design, the overall weight of the mount might be compromised, as elastomer materials often have high densities and, therefore, increased mass. The basic idea of the generated concept is to consider AM design principles of the shape-complexity-category (Table 2.14) to achieve vibration-decoupling due to integrated flexible geometrical elements into the mount housing. Those flexible elements should be purposefully designed according to NVH-simulation methods, considering shapes that might not be feasible with conventional manufacturing technologies. This way, specific eigenfrequencies could be diminished, resulting in a one-component design with an optimized trade-off between rigidity, vibration-comfort, and weight. With this approach, a minor component optimization could be identified, directly linked to the customer-relevant vibration and acoustic comfort requirement.

6.1.2   Experimental Approach Application

The feasibility of the presented approach for the customer-benefit-oriented application of AM potentials in Fig. 6.1 was tested in two online workshops (due to

COVID-19 pandemic), one in the academic and one in the corporate environment. The workshops were based on the use-case of a front section of a premium sports car as a reference vehicle. A total of 12 participants, mainly with expertise in automotive engineering, design methodology, and product development, were invited. The workshops were conducted according to the same procedure in both environments. This subsection summarizes the experimental application results. For further details compare [KFV23; Kus23].

The workshop concept consists of an introduction to DfAM, including purposeful and not effective examples of AM showcases to enhance the awareness of purposeful and less purposeful AM applications. The latter refers to additively manufactured applications with geometries designed for conventional manufacturing processes without rethinking the conceptual design. As positive examples, AM applications that are explicitly developed for AM were demonstrated, taking different DfAM potentials and limitations into account. Further, the workshop concept includes two interactive work sections (1st section compare Fig. 6.1 step 2 and 3, 2nd section step 4). The objective of the first phase is to utilize the provided method to identify conflicting objectives in the list of requirements or functional structure of the front end vehicle section.

Figure 6.2 shows the provided worksheets 1 and 2 as a simplified specification sheet, including a list of requirements, an abstracted functional structure, and an interactive 3D-overview of the conventional front end assembly of the reference vehicle. The list of requirements considers exemplary requirements in driving dynamics, comfort, consumption, and safety. The functional structure suggests exemplary high-level functional groups from each technical vehicle section (Fig. 2.20).

The conflicting requirements and functions identified by the participants were documented in the worksheet, prioritized and assigned to tangible involved assemblies or components of the reference vehicle (see Fig. 6.2). Several specific assemblies were identified in the two workshops with optimization potential linked to customer-perceived vehicle properties, e.g., wheel carriers, brake calipers, bumpers, and suspension components. Their AM suitability has also been rated according to the participants' expertise. In particular, large-volume shell parts such as bumpers were rated to have negligible AM suitability, primarily due to their size compared to available build volumes. After identifying conflicting objectives and assigning the respective assemblies, the first work phase was concluded, and the results were presented to the group.

The second work phase was initiated by introducing helpful DfAM tools. The tools suggested were the systematic network of AM design potentials [Kum18], the matrix of conflicting AM potentials [Fuc+20], and design rules for AM [Ada15]. The main task in this phase was to select up to 3 identified assemblies and generate alternative concepts for optimizing them with the help of the DfAM tools provided. The respective ideas were documented utilizing ideation sheets.

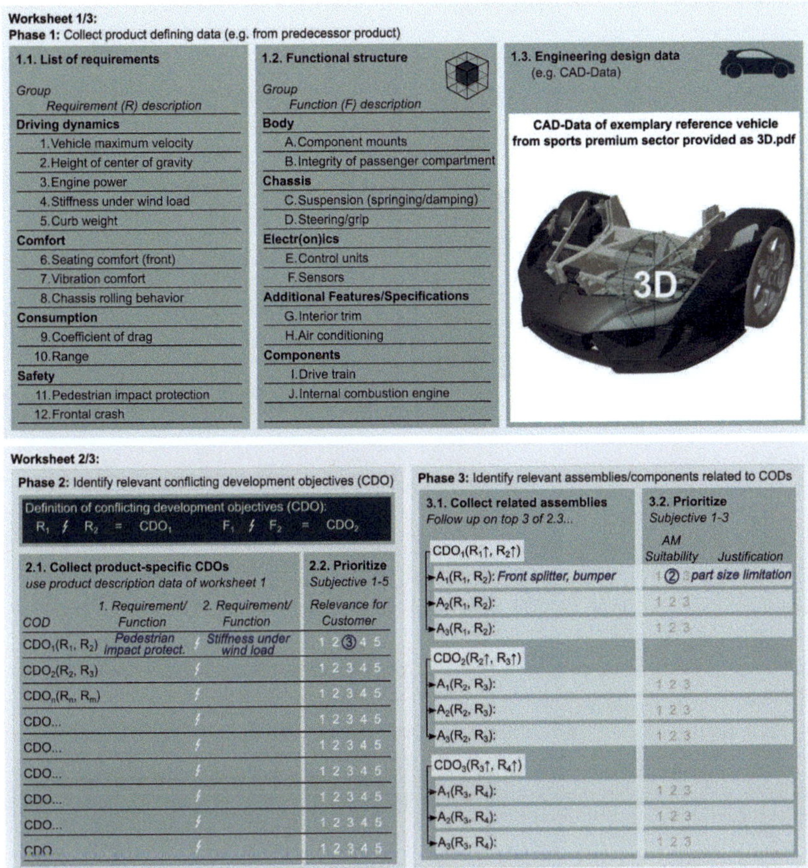

Fig. 6.2 Worksheet 1 and 2 to assist the identification of conflicting development objectives based on product defining data of an exemplary vehicle section

Figure 6.3 shows an exemplary sketch of a topology-optimized axle component for AM. The basic idea here is to achieve weight savings and reduced inertia of rotational masses through topology optimization. Relevant design principles applied from the systematic network of AM design potentials were, e.g., mesoscopic lattice structures, free-form surfaces, and functional integration, to achieve value propositions such as weight reduction, design space reduction, assembly cost reduction and reliability improvement.

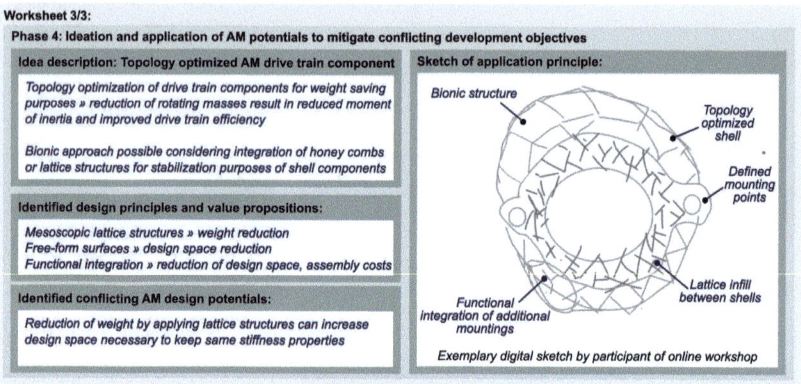

Fig. 6.3 Exemplary ideation sheet to specify the application of AM potentials according to identified conflicting objectives by workshop participants

Conflicting AM value propositions identified by the participants were design space reduction versus weight reduction, as a weight reduction often results in increased profile cross-sections. With the completion of the ideation sheets, the second work phase of the workshop was finalized and concluded with the presentation of the results to all participants.

Five distinguishable and comprehensible conceptional ideas were developed during the workshops, and two further ideas were sketched. Exemplary results addressed additively manufactured nodes for crash-relevant body structure, damping elements to mount the steering assembly, and topology-optimized axle components for the front end of the reference vehicle. In the context of these concepts, ten design principles and twelve value propositions from the systematic network of AM design potentials were applied, and five potential contradictions were identified between those using the matrix of conflicting AM potentials.

6.1.3   Summary and Discussion

One primary objective of this thesis is to explore and enable product-benefit-oriented application of AM design potentials. Therefore, several applications have been elaborated in the previous chapters accordingly. In this section, this central aspect of product-benefit orientation has been abstracted into a four-step approach to assist in identifying beneficial AM applications. The approach is focused on diminishing conflicting development objectives that often result in sub-optimal engineering trade-offs if based only on conventional manufacturing techniques. It is assumed that achieving optimized trade-offs between conflicting objectives due to additional design freedom can result in significant product benefits and, therefore, customer-relevant product improvements. This approach ensures that each application identified can be traced back to customer-relevant trade-offs in the respective product properties. Further, this section describes the four main steps of the approach in detail. Additionally, for experimentation purposes, this section covers how the approach has been applied in two interactive online expert workshops in the academic and corporate environments. Furthermore, the DfAM tools suggested by the approach were utilized, and their strengths and weaknesses were identified accordingly.

In conclusion, the proposed approach for product-benefit-oriented application of AM potentials represents an opportunity to examine existing and new products for AM-related optimization potentials, including the direct relation to customer-relevant product properties. According to the evaluation, the participants rated both the approach and the tools used as valuable and helpful for the purposeful application of AM potentials. Comprehensible concepts were developed in the course of both workshops, which, in addition to the AM potentials identified, also highlighted some conflicts among them. However, those could be mitigated with the help of the matrix of conflicting AM potentials (Sect. 3.3). Additionally, the conceptual ideas generated in the academic and corporate contexts were similar, so no significant systematic differences in the results could be distinguished. One reason might be that participants were experienced in the field of AM, or the scope of the front end of the vehicle of reference only allows for a limited number of ideas.

The evaluation of the workshops revealed that the workshop concept, particularly the structure of the work phases, and the successful generation of own ideas were approved by the participants. Most participants were satisfied with the workshop results and rated the suggested approach as purposeful. Regarding the applied methods and tools, the systematic network of AM design potentials [Kum18], the matrix of conflicting AM potentials [Fuc+20] and their use was assessed as purposeful and useful. A limitation of the overall approach is the applicability to products of high complexity, such as an application on a complete vehicle level. In such a case, one conflict in objectives affects a multitude of assemblies and parts, which again might also be relevant to several other conflicting objectives. Here, additional complexity management methods can be helpful.

Although this work addresses some challenges in DfAM, there is still a need for action in this field of research to ensure a product-benefit-oriented application of AM potentials and, thus, to fully leverage the optimization potential offered by this technology. For example, an additional research focus concerning DfAM is complexity management in the context of engineering design itself. The complex geometries enabled by AM do not necessarily increase manufacturing efforts significantly. Moreover, complexity increases in the early phases of PDP. This increased complexity, necessary to realize complex and highly optimized designs (e.g., bionic, functionally integrated, lightweight structures), challenges conventional CAD and CAE tools, making the engineering design process significantly more time-consuming. This aspect is further elaborated in the following sections Sect. 6.2 and Sect. 6.3. Concluding, enabling additional and otherwise not feasible customer value in upcoming products is one central pillar to propel AM technologies forward. The approach presented in this section might contribute accordingly.

6.2   Necessary Advances in Computer-Aided Design Tools

Engineering design is considered purposeful when required product functions are realized, considering the technical and economic feasibility of the company's resources. To do so, complementary to CAD, additional *Computer-Aided Engineering* (CAE) tools became crucial to generate, simulate, and optimize complex designs. Nowadays, CAD and CAE tools are well aligned with conventional manufacturing processes (e.g., milling, casting, molding). However, upcoming manufacturing technologies, such as AM, differ significantly in their restrictions and, thus in the design freedom from conventional manufacturing processes. The holistic utilization of AM design potentials, relying on conventional CAD tools and workflows, is therefore impeded [Fuc+22].

This section is focused on the limitations of state-of-the-art CAD mentioned in the literature and encountered during the AM-related design elaborations of this work, particularly in Chap. 4 and 5. The following content is based on the CAD-fundamentals introduced in Sect. 2.1.2. The following Sect. 6.2.1 summarizes DfAM-related requirements on CAD systems that still need to be addressed by standard commercially available CAD software tools. In a subsequent step, Sect. 6.2.2 projects and locates those general improvement points in the structure and composition of CAD software tools. By doing so, a contribution to the overall objective of enabling design engineers to leverage AM design potentials holistically is expected.

For a more detailed elaboration on this topic, consult the work of FUCHS ET AL. (2022) [Fuc+22].

6.2.1   DfAM-Related Requirements on CAD

One main objective of the DfAM field of research is to enable design engineers to leverage AM potentials [Gib+21; Tho+16; KWV16]. As AM processes reach technological maturity in many industries, engineers aware of DfAM are increasingly struggling to exploit new design potentials depending on conventional CAD tools. For this reason, alternative software solutions are emerging. Those alternative tools address AM-related pain points, facilitating product engineering design and optimization regarding their topology, geometry, functional architecture, and manufacturability. The main objective of this subsection is to address the need for further development in engineering design tools to cope with increasing AM-related design complexity. For this purpose, yet-to-be-addressed DfAM-related requirements on state-of-the-art CAD tools are gathered from the literature, structured, and complemented.

Even though highly optimized and often complex geometries can be manufactured with AM, modeling such designs in conventional CAD tools can be challenging. One reason is that CAD tools have evolved continuously along conventional manufacturing processes over the past decades. Therefore, design challenges related to conventional manufacturing processes were addressed during CAD development. As a result, essential functions for the design of high-complexity parts still need to be in development focus or have only been rudimentarily implemented.

This demand leads to several alternative software tools or plug-ins emerging in the software landscape, offering supplementary functional scope compared to conventional and established CAD tools. Table 6.1 complements the listing of Table 2.6 with emerging alternative software tools. Here, it is essential to consider that 3D modeling software focused on computer-aided-styling (CAS) might not be suitable for accurate technical engineering design (e.g., not mathematically accurate, no design history, no associativity, no reproducible results). Nevertheless, in the example of Grasshopper, a plug-in for the CAS software Rhinoceros, it becomes clear that CAS-tools are also being used for complex DfAM design challenges (compare Fig. 4.3 for exemplary application).

In Table 6.1, alternative software solutions are mentioned that emerged mainly during the last decade, offering additional benefits to engineering design in the DfAM context. Those examples are either alienated from other neighboring design sectors to mechanical engineering (market pull), e.g., architectural design, or being purposely developed for and promoted in the DfAM community (technological push). To name a few, the additional benefits enabled by those alternative tools range from increased stability, the agility of parameterized complex geometry representations, graded material properties, or voxel-based generative design functions up to enhanced processing performance (see Table 6.1 *particular feature*).

Table 6.1 Exemplary overview of emerging software solutions addressing DfAM-related pain points in CAD based on [Fuc+22] complementary to Table 2.6

Category		
Software product	*Company*	*Particular feature*
Emerging software beneficial for DfAM (partially commercially available)		
nTopology	nTopology	agile geometry representation, field driven design
Dyndrite	Dyndrite	performance (GPU), agile geometry representation
Grasshopper	Scott Davidson	parametric visual scripting
xGenerative Design	Dassault Syst.	parametric visual scripting
Dynamo Studio	Autodesk	parametric visual scripting
Netfabb		graded mesoscopic structures
Monolith (*discontinued*)		voxel-based generative design
Frustum	PTC	generative design
Hyperganic	Hyperganic Group GmbH	parametric generative design
Synera (fka. Elise)	Synera GmbH	bionic/algorithmic generative design, parametric visual scripting

Further, the literature also points out necessary advancements in CAD software tools for DfAM. By gathering those needs in related work and complementing them with requirements gathered from applications in this thesis, at least six categories relevant to the design workflow can be distinguished (Fig. 6.4). Some requirements address shortcomings in the overarching *design-workflow and product structure* itself. The remaining five categories point out specific requirements, e.g., necessary functionalities in *material* and *geometry representation* in the CAD system. Also, requirements regarding the *geometry manipulation* by design engineers themselves are in focus. Additional requirements addressing the CAE context, e.g., *simulation* and *geometry optimization/generation*, are also addressed, as these steps are crucial to achieving highly optimized yet feasible DfAM designs. In the following paragraphs, exemplary requirements from each of those major categories are further clarified.

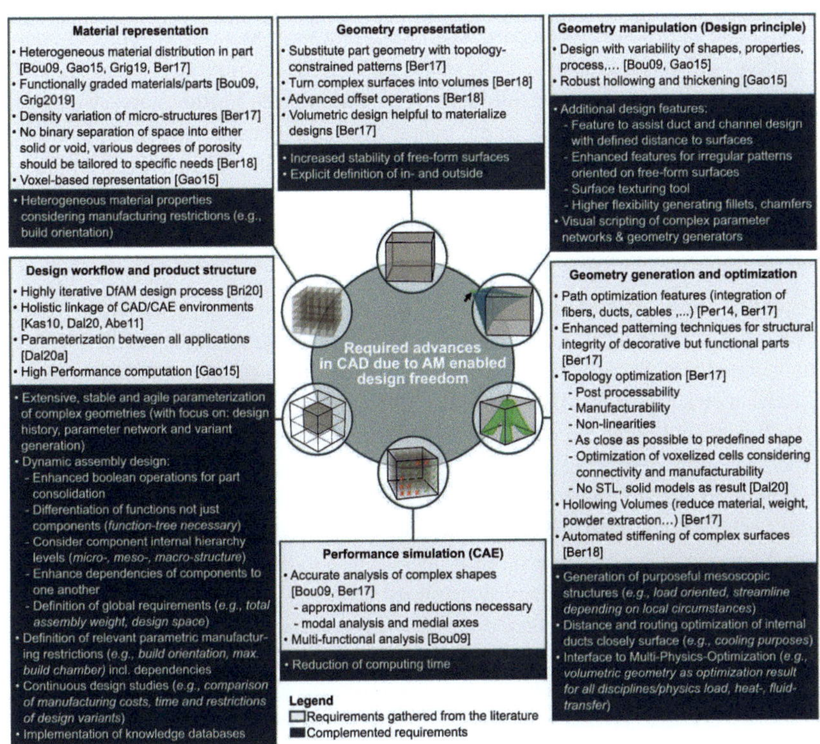

Fig. 6.4 Overview of necessary advancements in computer-aided design tool development gathered from the literature and complemented by FUCHS ET AL. (2022) [Fuc+22]

Material representation

Most needs expressed in this category are derived from the capability of several AM processes to influence material properties or interchange different materials [Gib+21; Tho+16] in a voxel-based manner [Gao+15; Kum18]. First, the possibility to vary material properties, e.g., a graded density/porosity, inside one part geometry would enable the design engineer to tune the material's performance to differing local requirements. This aspect should be possible to design on different hierarchical levels, e.g., micro- (e.g. grain structure of metals [BFR17]), meso- (e.g. cellular structures like lattices and honeycombs [BFR17]) and macroscopic- (e.g. differing transitioning materials in certain areas of the part [Gri+19]) level (compare Sect. 2.3.3). Nowadays, this is often carried out with the workaround of designing stand-alone partitions of one part. Then, different materials are assigned to each partition before printing. However, gradients/transitions in property and kind of material are not yet feasible following this workaround. Furthermore, heterogeneous material properties due to the manufacturing process (e.g., tensile strength weakness perpendicular to layer) should be considered in the CAD/CAE environment.

Geometry representation

Regarding the geometry representation in CAD environments, additional needs result from challenges in the transformation of optimization results (often triangulated/tessellated hollow surfaces, e.g., STL) into editable CAD data (e.g., V-Reps or closed B-Reps) [BHD18] (Fig. 6.4). This need is already being addressed by software tools such as, e.g., *LEOPARD* and *3DExperience*, that smoothen and export topology optimization results in the subdivision surface format (Table 2.5), enabling this way to finalize the design in the CAD environment. Strongly dependent on the kind of geometry representation, e.g., V-Rep or B-Rep (Table 2.5), are also functions such as advanced offset operations to hollow out volumes [BHD18], substitute surfaces with topology-constrained patterns [BFR17] and materialization of geometries [BFR17]. Considering that each approach to geometry representation suits specific applications more than others, enhanced interchangeability of the representation approaches is necessary. The *Dyndrite* software is one example that addresses this need, considering the correct type of geometry representation for each design task (see Table 6.1).

Geometry manipulation (Design Principle)

Apart from general design challenges, e.g., the high number of involved software tools and the iterative character of the DfAM processes, the hands-on engineering design workflow is not yet addressed in the literature. In complex AM applications, the hands-on DP, particularly managing design and simulation complexity, is the bottleneck rather than the production complexity itself. Thus, in addition to conventional (compare Sect. 2.1.2), new design features and geometry manipulation principles are necessary to utilize AM design potentials further. e.g., additional features, such as assisted duct and channel design and tools to assist in designing irregular patterns and textures on freeform surfaces, would significantly simplify laborious modeling problems (Fig. 6.4).

Besides additional features in conventional CAD environments, certain complications cannot be handled with the conventional direct modeling technique. One example is based on the tool-less character of AM technologies, which enables design variations, improvements, or adaptions with little impact on development costs compared to tool-based manufacturing techniques. Nonetheless, a new level of design parametrization is required to generate and alternate designs with feasible effort. CAD data should be comprehended less as a virtual adaptive 3D model targeting a design freeze and more as parametric geometry generators, which, based on adaptive input parameters, repeatedly deliver updated geometries based on the same underlying concept. For this purpose, additional design principles such as *visual scripting* environments should be implemented in conventional CAD tools, allowing design engineers to program their geometries easily. As examples, *xGenerative Design* (*3DExperience* platform) and *Synera* are implementing visual scripting into the technical engineering-focused CAD environment (Table 6.1).

Geometry generation and optimization

As AM capabilities can exceed human hands-on modeling capabilities, algorithmic geometry generation (including topology optimization) is frequently applied in the DfAM context. Shortcomings of optimization software mentioned in the literature either address missing functionalities or suggest improvement of already existing ones (see Fig. 6.4). Regarding topology optimization, the integration of manufacturability and post-processing restrictions, a V-Rep export of the resulting blueprint geometry ready for detailing, and optimization functionalities on the voxel level are pointed out as development tendencies by the literature [BFR17]. Furthermore, independent of the AM context, the need for topology optimization to consider non-linear plastic material behavior is crucial for implementing the method for design challenges focusing on energy dissipation through material deformation, such as crash-relevant structural vehicle parts. Additional functionalities mentioned and desired by the field of research are path optimization tools for integrated fibers and ducts (e.g., Fig. 4.13 based on [BBM22]), enhanced structural patterning, hollowing techniques, and automated stiffening of complex surfaces [BFR17; BHD18]. Moreover, generating purposeful mesoscopic structures (e.g., truly load-oriented structural lattices), optimization of distance and route of component internal ducts (e.g., close to freeform surfaces), and multi-physics optimization capabilities are additional areas of improvement for upcoming optimization tools.

Performance simulation (CAE)

One major challenge that design engineers confront during performance simulation of complex geometries is the processing capability of CAE systems [BFR17]. BERMANO ET AL. (2017) indicate that often approximations or reductions are necessary to obtain computable tessellated models with a reasonable amount of resources. Those simplifications can result in distrust of the significance of simulation results and, subsequently, the part's performance in real use. Furthermore, due to AM technologies' functional integration and part consolidation potentials, parts optimized for AM can take over an increased functional scope compared to conventional designs. This circumstance leads to multi-functional complexity of boundary conditions that need to be modeled in future simulation environments [BR09].

Design workflow and product structure

General shortcomings towards the number of different software environments necessary for DfAM and the missing cross-cutting parametrization [DPL20], the resulting highly iterative workflow [BSZ20] and the limited compatibility of CAD and CAE environments [KBF05; AR11; DPL20] are pointed out in the literature. Also, the overall need for additional computation performance is expressed [Gao+15] accordingly (Fig. 6.4). Additionally, specific needs towards stable, parameterized, complex geometries with design history enabling the generation of multiple variants have been identified. Here, visual scripting software such as, e.g., *Grasshopper, xGenerative Design, Dynamo Studio* are being used for the development of geometry generators that enable to leverage individualization potentials of AM, realizing applications like individualized products lot size one (compare Table 6.1 and Fig. 4.3). Furthermore, dynamic assembly design features would be helpful to support functional integration and part consolidation potentials of AM (Fig. 6.4). Here, a function-centered product structure is crucial in addition to the part-centered. Further, the possibility of setting AM restrictions on the assembly level, carrying out AM design studies (as already partially realized in the 3DExperience platform), and implementing DfAM knowledge databases should also be considered in the future.

6.2.2 Development Areas in CAD-Architecture

Analyzing the six DfAM requirement categories on CAD, explained in the subsection above, and interlinking them with the basic CAD architecture (Fig. 2.10) leads to the identification of at least four major development areas in the architecture of CAD-Systems. These areas are illustrated in Fig. 6.5 and explained in more detail.

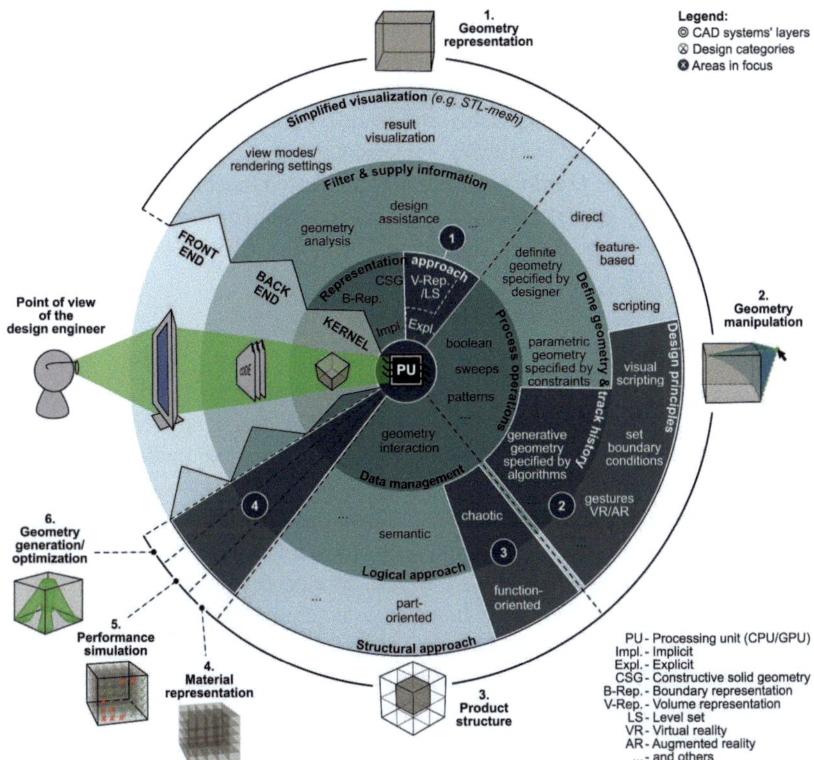

Fig. 6.5 Identified focus areas for further CAD-system development based on [Fuc+22]

1. Multiple Geometry Representations:

As introduced in Sect. 2.1.2, several mathematical boundary- or volume-based approaches can be considered for geometry representation in CAD kernels (Fig. 2.10). CAD systems are only based on particular representation approaches (mostly *implicit* methods), often leading to conversion efforts or dead-ends (e.g., STL-Files) further down the line. However, none of those approaches can be indicated as the single most suited approach to complex AM designs for the future. One central pain point of DfAM is missing capabilities to represent heterogeneous material distributions as well as micro-, meso- and macroscopic non-uniform structures inside a part. Further, CAE functionalities are insufficient to design those structures for a specific use case purposely. Therefore, development efforts should focus on the capabilities of upcoming CAD systems to support, convert, and combine implicit as well as explicit representation methods using the right approach for the right design task (compare area 1 in Fig. 6.5). The main challenge here might be transitioning from one approach to another stably and seamlessly without losing information. Alternative software tools are already implementing similar functionalities to such an extent (see example *Dyndrite* in Table 6.1). Enhanced functionalities in this area will simultaneously improve several DfAM-related pain points, e.g., performance-related issues during the design, simulation, and optimization of complex geometries.

2. Versatile Design Principles:

Today, design engineers are the central and fundamental element of the design process [Pah+07]. This constellation also applies to future design problems. However, in the DfAM context, engineers can be a limiting factor while modeling complex geometries. If geometry manipulation functionalities of CAD tools are not effortless enough, design simplifications occur as shortcuts to reduce modeling efforts, resulting in part performance losses. In such cases, it is crucial for design tools to facilitate complex design tasks further. Hence, the category *geometry manipulation* sets focus on different manners in how design engineers are enabled to model or generate virtual designs via the front end of the system (see Fig. 2.10). As introduced in section 2.1.2, nowadays, most CAD solutions are based on a combination of the design principles *direct*, *feature-based*, and sometimes *script-based* modeling approaches. These design principles are laborious and inefficient in achieving complex geometries. Additional design principles, e.g., *visual scripting*, are emerging among complexity- and variant-driven designs. *Visual scripting* enables design engineers to program associative parametric geometries by interlinking functional building blocks with in- and output parameters in a logical diagram. The resulting functional diagrams represent geometry generators, in which parameter adjustments directly lead to new design variants in an agile, stable, and manageable process. This design principle is a crucial feature of emerging software solutions (compare *Grasshopper, Dynamo Studio, xGenerative Design* in Table 6.1). It is often used in specification tools for individualized products or for generating algorithm-based hierarchical geometries.

Due to the limitation of humans to determine highly optimized geometries and the intensive labor necessary to directly design them, *algorithmic geometry generation* plays an increasingly important role in the DfAM context (e.g., topology optimization, generation of mesoscopic structures). As mentioned in Fig. 2.10 *set boundary conditions* is the crucial step for algorithms to generate purposeful geometries within a characterized design space. Setting correct boundary conditions for specific use cases is an engineering task that generative design tools cannot yet conduct. Therefore, it will increasingly be part of a design engineer's workflow. Thus, *set boundary conditions* can be considered an additional design principle and should be one focus of geometry manipulation functionalities in upcoming CAD systems. This relation emphasizes the ongoing aggregation of CAD and CAE tools. Similar to the different geometry representation approaches mentioned above, each design principle has advantages for a specific design task. The seamless combination of those design principles should also be addressed in the development of future design tools (see, e.g., *xGenerative Design* in Table 6.1).

These and further design principles such as *gesture-based virtual/augmented reality* bear the potential to increase manageable design complexity and are considered the principle second development area for upcoming CAD systems (compare area 2 in Fig. 6.5).

3. Function-oriented product structure in front- and back-end:

Conventionally, the product structuring approach implemented in CAD environments is based on part-assembly-affiliations. This structuring approach makes sense as conventional complex products comprise numerous individual parts. However, an assembly structure does not necessarily resemble the functional structure of a product, as different parts might be necessary to fulfill one function or vice versa [Pah+07]. Particularly in the DfAM context where *part consolidation* and *functional integration* are considered AM design potentials, a product's assembly and function structure might differ intensively. DfAM applications often show part count reduction and decreased assembling complexity. Nevertheless, upcoming CAD systems need to address increased functional complexity on the part level. Even though functionally complex and merged parts might be feasible with AM, component boundaries must be chosen wisely and can shift iteratively during the design process.

Additionally, specific functions must be interlinked to geometrical features or part segments to set localized and variable material properties, enabling optimized parts and their performance simulation. Consequently, *function-oriented* product structures should be implemented in future CAD systems in addition to *part-oriented* structures. This approach would enhance dynamic assembly design functionalities such as alterable partitions between components and, e.g., advanced boolean operations, assisting design engineers to apply DfAM functional integration potential. This aspect concludes the third area for further development in CAD (compare area 3 in Fig. 6.5).

4. Enhanced CAE functionalities:

The fourth area for further CAD development summarizes necessary improvements regarding material representation functionalities, enhanced simulation performance, and CAD-native optimization procedures (compare area 4 in Fig. 6.5). This area enhances today's CAD scope towards implementing and extending CAE functionalities. Besides the additional necessary AM-specific functionalities, this development tendency implies that design and simulation engineers must work closer to achieve optimized designs. In other words, the activity scope of design and simulation engineers will be less distinguishable.

6.2.3   Summary and Discussion

The main objective of this section is to address the need for further development in engineering design tools (CAD/CAE) to cope with increasing design complexity and thereby enable design engineers to leverage AM design potentials. A general finding of this section is that numerous yet-to-be-addressed requirements of DfAM on CAD exist mentioned in literature and gathered from applications in this thesis (Fig. 6.4). Furthermore, it has been shown that the development of additional CAD features is insufficient to address those needs. All four main layers of CAD systems (*front-end, back-end, kernel* and *processing unit*) require further development and increased versatility (Fig. 2.10).

The outcomes of this section are the overview of the CAD characteristics from a design engineer's perspective (Fig. 2.10) and a structured collection of DfAM-related requirements on CAD (Fig. 6.4). Additionally, a tabular overview of exemplary alternative software solutions addressing particular DfAM needs is provided (Fig. 6.1). Based on those findings, four major areas for development in the CAD structure were identified. Addressing these four development areas will enable design engineers to overcome design challenges with increased complexity more efficiently.

Implications for future research focusing on CAD systems consist of further investigations of back-end related consequences and measures necessary to realize DfAM requirements pointed out in this work. With a focus on the software landscape, a detailed analysis of DfAM requirements not yet addressed by alternative software solutions would provide further insights on development white spots for software developers. Another open aspect to further analysis in product development are valuable methods to manage requirement complexity and conflicting objectives. Also, comparing increased product reliability by minimizing points of weakness in design (e.g., intersections) as opposed to decreased ease-of-repair and increased cost-of-repair due to part consolidation and functional integration should be analyzed in more detail.

In conclusion, a contribution towards the convergence of methodological DfAM and CAD development is provided within this work, enabling a scientific discourse with the potential to improve engineering methods and tools further.

6.3   Engineering Design Workflow Fit for DfAM

To fully leverage AM design potentials, it is necessary not only that CAD/CAE environments are further developed, as stressed in Sect. 6.2, but also that the hands-on design workflow, from a design engineer's perspective, needs to be adapted. This aspect is clarified in this section.

DfAM propels highly optimized designs further as AM technologies shift conventional manufacturability restrictions. As a result, AM-related manufacturing complexity decreases, as no additional tools with multiple parts and functional surfaces need to be engineered. However, part-related design complexity can increase significantly for design engineers. Additionally, independent of the manufacturing process a part is designed for, the limited resources and lack of design time motivate design engineers to take shortcuts during the design workflow. For example, edges and corners are easier and faster to design accurately than freeform surfaces, independently of the fact that they are more straightforward, e.g., to mill too. With this in mind, if the design workflow stays cumbersome and complex, design engineers will not be able to utilize AM-related design potentials fully.

Contemporary work focuses on implementing DfAM-related design potentials and restrictions as a general methodological framework [KWV16; Sch+20; Li+21], provide DfAM-related knowledge for design engineers [Ada15; VDI21a; Kus23], and implementing and automating DfAM design restrictions in CAD-Systems [Wib21]. So far, no hands-on design workflow has been found in the literature that enables design engineers to generate and manipulate functionally complex and highly optimized CAD geometries.

This section elaborates on a workflow necessary to enable complex geometries and to counteract this step to be the bottleneck of the DfAM development process. The engineering design workflow proposed in this section is based on the MDO vehicle segment application conceptualized in Sect. 4.2.2. Here, a high level of functional and geometrical complexity results from the optimization-based concept. In the following, the approach applied to the MDO vehicle segment is first generalized in Sect. 6.3.1 as a multiphysics design approach based on design space sensitivities. In Sect. 6.3.2 the generalized approach is implemented in the hands-on conventional engineering design workflow already introduced in Fig. 2.11. For a more detailed elaboration of this topic, compare FUCHS ET AL. (2020) and FUCHS ET AL. (2022) [FHK20; Fuc+22].

6.3.1   General Approach for Multidisciplinary Design Space Optimization

The elaborated multidisciplinary vehicle segment conception described in Sect. 4.2.2 strives towards realizing an optimal interdisciplinary compromise between computationally generated locally optimized geometries for each relevant functional category of the application. Thereby, the approach intentionally disregards manufacturing restrictions at first. The obtained results are later analyzed considering DfAM-related potentials and restrictions. This procedure sets the fundamental characteristic of the generalized approach described below. This characteristic implies three assumptions:

1. Finding the best global (holistic) compromise between all product functions has a higher beneficial impact on the product's overall performance than finding local (individual) optima for certain customer-relevant functions, thereby neglecting others.
2. Considering the manufacturing technologies with the least amount of design restrictions enables the closest design solution to computationally optimized results and, therefore, achieves better product performance than more restrictive manufacturing technologies.
3. Less geometrically restrictive manufacturing technologies, e.g., AM, tend to be cost-intensive and should only be considered where design complexity benefits efficient function fulfillment.

The following approach does not claim to cover the DfAM-focused DP holistically. It should only be considered as a substep of the overall DP. Considering the DP based on [VDI87; VDI19b; VDI04] summarized in Fig. 2.5, and the DfAM framework adapted from the VDI-based process by KUMKE ET AL. (2016) [KWV16, p. 12] this approach becomes relevant during *phase II, conceptual design* after *step II, determine functions and their structures* where the function structure of the product is already defined and *step III, search for solutions principles and their combinations* starts. At this stage of the DP, the technical package might be roughly outlined in its central maximum and minimal dimensions, and the first modules are distinguished and located, resulting in a first rough definition of the design space available for the product. As this approach relies on computationally generated geometry blueprints, there is no definitive differentiation between CAD and CAE (compare Fig. 2.9 for differentiation) as elements of both areas interfuse to realize optimized designs. Fig. 6.6 illustrates the basic five-step design approach on the top.

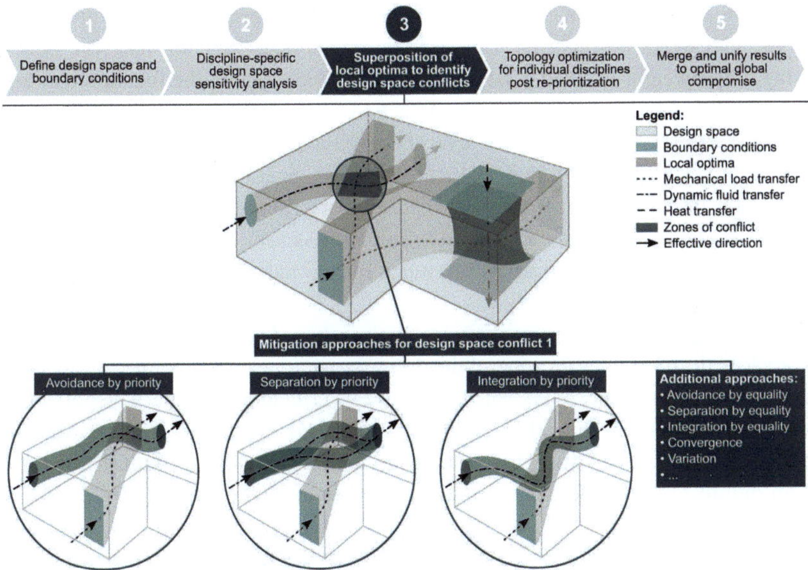

Fig. 6.6 Approach for a interdisciplinary optimization of design spaces based on the identification of conflict zones between local optima with focus on conflict mitigation principles

With a conventional conception approach, the available design space is often divided strictly between components and functions according to their priority regarding safety relevance and indispensability. However, this procedure counteracts functional integration potentials. Figure 6.6 illustrates in the center an outer shell that defines the available design space based on the technical requirements of a component or product. Further, boundary conditions consisting of physical intersections necessary for multiple functions to be realized within the design space are described by surfaces on the outer shell. In this case, static load, dynamic mass, and conductive heat transfer are exemplary, indicated by vectors on the center of each intersection surface. The direction of the indicating vectors distinguishes between in- and output interfaces. These boundary conditions are derived either from product requirements, function structure, technical package, or based on intersections to neighboring components in the assembly.

The setup illustrated in Fig. 6.6 in the center implies that differing volumetric areas of the design space are more or less relevant for each technical discipline. Even though not yet quantified in their dimensions, these areas become apparent by interconnecting the related in- and output interfaces most shortly and efficiently independent from each other. The shorter the interconnection of interfaces is, the less material might be necessary to embody the function later on, benefiting a lightweight-focused concept. However, depending on the function and the optimization target, the shortest interconnection might not be the most suitable solution. For example, in the physics area of fluid dynamics, short yet not optimized media ducts can lead to turbulence flow and high-pressure losses. In contrast, a smoothly curved yet longer duct might cause less pressure losses and turn out more energy efficient. Therefore, additional information is necessary to characterize the design space further regarding its relevance for the individual functions, physical categories, or technical disciplines.

A subsequent design space sensitivity analysis based on CAE methods is helpful to enclose the most relevant areas of the design space for each discipline. A superposition of the individual results potentially reveals critical design space disputations between functions, technical disciplines, or physical categories. If not solved by re-prioritization of one discipline or another, these conflict zones indicate areas of high multidisciplinary functional integration potential (compare Sect. 4.2.2 for a tangible application).

Figure 6.6 illustrates on the bottom three exemplary re-prioritization approaches for design space conflict mitigation. Those approaches are highly dependent on the compatibility of physic categories. The necessity of a particular function for material or void in that specific zone of conflict is thereby one of the main incompatibilities. Next, the specific requirements for the material properties might differ between functions. *Avoidance by priority* refers to the most conventional approach. The less critical function, in this case, the media duct, has to evade the zone of conflict with the load-bearing strut. *Separation by priority* follows the same principle, yet also suggests a division of the less prioritized function. *Integration by priority*, however, points out that integrating the lower prioritized function into the higher might lead to a more design space-efficient solution. In this example, the conflict of fluid dynamics with mechanics, the neutral axis of the load-bearing member might also be utilized to integrate a media duct, as the material here does not contribute significantly to load transfer. Even though, as a consequence, the load-bearing cross-section might need to expand slightly due to the duct integration, the overall result might still be more efficient regarding the usage of design space and material.

These mitigation approaches for design space conflicts are only three examples. Figure 6.6 also lists on the bottom right further approaches that might be useful when prioritization of functions is not possible. Furthermore, considering DfAM design potentials, particularly for those conflict zones, further expand possible solution space, e.g., by considering geometrical, functional, hierarchical, and material complexity (compare Table 2.14). When the multidisciplinary design space exploration and characterization is concluded, and the necessary solution principles for design space conflict mitigation are defined, concept detailing begins. At this point, component partitions, material categories, and manufacturing processes are elaborated. This way, the detailed optimization of each component's geometry considering available design space, manufacturing restrictions, and functional integration is enabled.

Figure 6.7 summarizes the five-step multidisciplinary approach described above, considering four physical disciplines in a swim lane configuration. For each discipline, exemplary boundary conditions and optimization targets are listed. Each lane demands specific CAE methods for sensitivity analysis (*step 2*) and topology optimization (*step 4*). Future MDO software tools might also support the superposition of optimization results (*steps 3 and 5*). Due to the limitations of CAD and CAE tools, design engineers are challenged to accomplish those design steps manually for now.

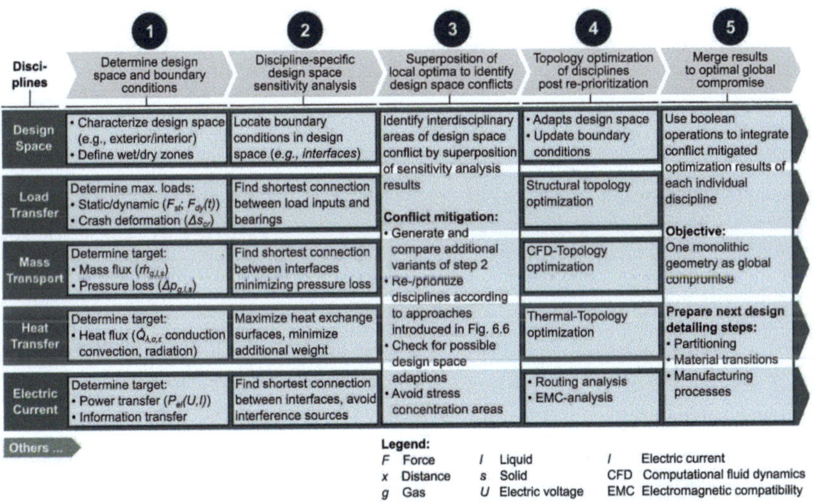

Fig. 6.7 Five step approach for an interdisciplinary optimization of design spaces

6.3.2   Necessary Adaptions to the CAD Workflow

This subsection elaborates on a visionary hands-on design workflow inspired by DfAM based on the conventional CAD workflow. The adaptions are derived from the approach to multidisciplinary design space optimization introduced in Sect. 6.3.1 above.

Adaptions to the hands-on engineering design workflow are considered necessary in the context of DfAM to cope with increased product complexity. These adaptions to the design workflow follow the objective of implementing FEM-based multidisciplinary optimization of topology and geometry as central design elements. This aspect is motivated by the capabilities of AM technologies in the realization of optimized geometries with only a few adaptions due to DfAM manufacturing restrictions.

The following descriptions refer to an envisioned design workflow described in Fig. 6.8 (*right*) and can be compared with the exemplary conventional design workflow in Fig. 6.8 (*left*) derived from Fig. 2.11. The schematic diagram is focused on a new design. Partial aspects might also apply to a re- or variant design. The diagram further includes a description (*D*) and the respective output (*O*) of each significant step. Bidirectional arrows imply the iterative character of design tasks.

The first significant difference between contemporary and prospective design workflow identified is the increased emphasis on the component/assembly level of the process. The detailed part design (*step 3*) is only approached after sensitivity analysis of the design space (*step 2A*) followed by a rough embodiment design phase (*step 2B*), both of which are conducted still on assembly level. The second differentiation is the emphasis on the functional composition of the component prior to the geometric design [Fuc+22].

The main structure of the envisioned workflow of Fig. 6.8 is based on the following three assumptions [Fuc+22]:

1st Assumption: Due to AM design freedom, AM components are prone to reach higher levels of functional complexity in comparison to conventional manufacturing processes.

2nd Assumption: With an increased amount of functions to be realized in one component, a higher probability of design space disputations in between those functions can be expected.

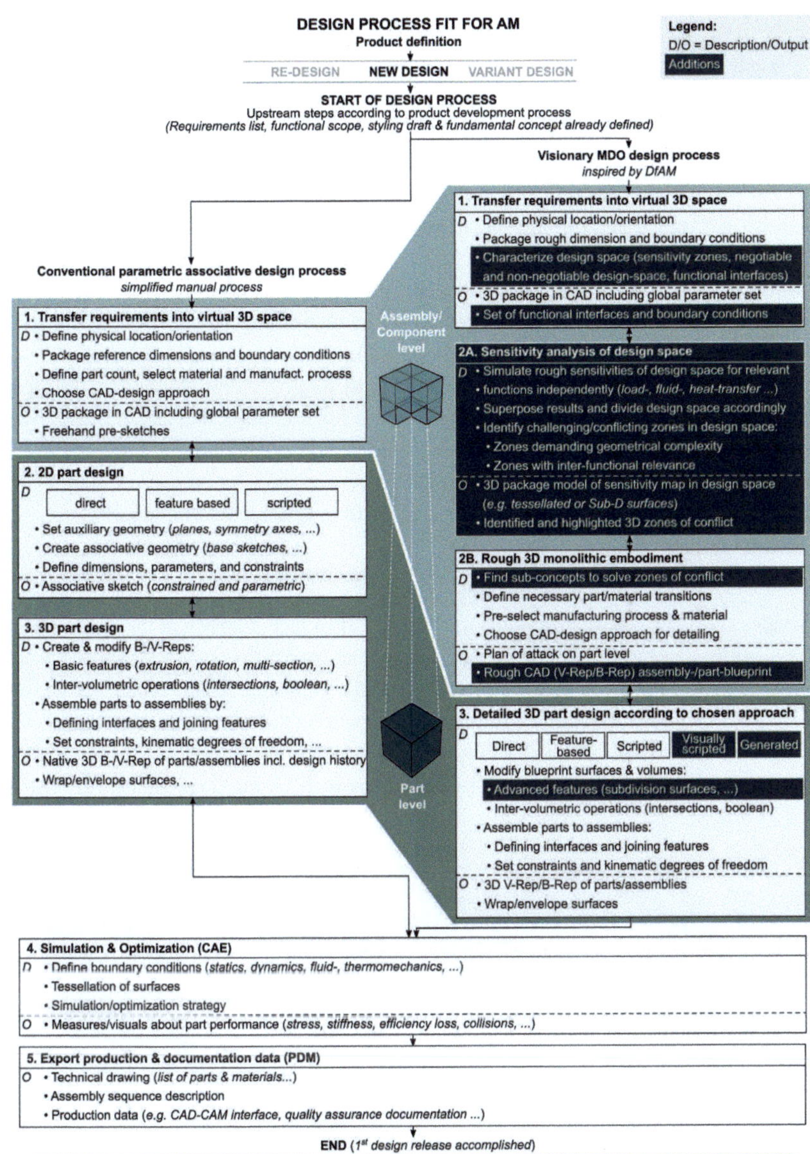

Fig. 6.8 Step-by-step conventional parametric-associative CAD workflow (left) compared to suggested visionary engineering design workflow fit for DfAM (right) based on [Fuc+22, p. 9] and Fig. 2.11

3rd Assumption: In comparison to conventional manufacturing technologies, AM is often considered to manufacture computer-generated optimized geometries that exceed human design capabilities, leading to increasing utilization of generative algorithm-based design principles in the DfAM context (e.g., topology optimization, generation of mesoscopic structures)

Based on these assumptions, the upcoming design workflow should cope with increased geometric-, hierarchical-, material-, and function-related complexity and further support the management of design space disputations in between functions. If the set of functions to be implemented in the component of choice is assignable to more than one technical section (compare Fig. 2.20), the term *multi-disciplinary design* or more than one physics faculty (e.g., mechanics, thermodynamics) the term *multi-physics design* is considered appropriate for this particular design task. The following step-by-step additions or adaptions to the conventional design workflow are envisioned as necessary to manage the increased design complexity mentioned above [Fuc+22] (compare Fig. 6.8 for a high-level approach summary):

Step 1: Transmission of requirements into virtual 3D space:

Besides transmitting requirements into virtual 3D space, already in the first step of the workflow, a more detailed design space characterization is necessary to set up the foundation for the following sensitivity analysis of the design space. As generalized in Fig. 6.9, a detailed design space characterization includes, among others, e.g., boundary conditions of all kinds, definitions of safety-critical zones, negotiable and non-negotiable design space limitations as well as interface specifications, if

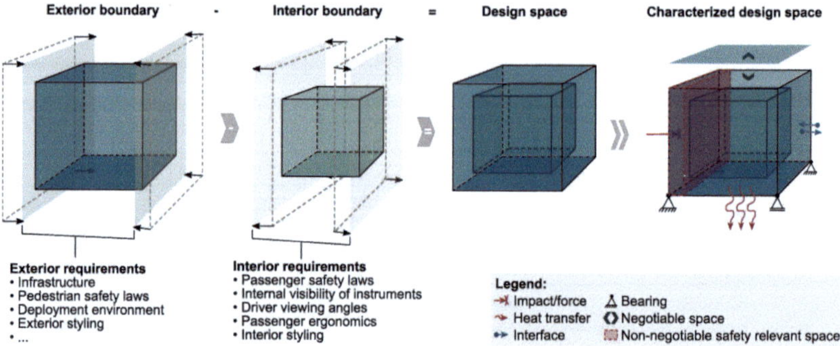

Fig. 6.9 Exemplary generalized design space characterization derived from automotive-relevant exterior and interior requirements

already determinable. This proceeding means that after the first step, the designer engineer does not proceed with manually designing auxiliary geometries (mostly in 2D) necessary for the 3D part or shape design (compare Fig. 6.8 left step 2). Instead, based on the characterized design space, the set of boundary conditions, and necessary CAE functionalities, the design engineer carries out a *sensitivity analysis* (step 2.A) of the design space for each function.

Step 2A: Sensitivity analysis of design space:

The main objective is to determine the most relevant segment of the available design space to realize each component function to its optimum (e.g., functions similar to load-, fluid-, or heat-transfer). Nowadays, this step often happens primarily based on the cognitive experience and intuition of the design engineer, which might be insufficient for design tasks with increased function and geometric complexity. The sensitivity analysis envisioned here relies on early-stage simulation functionalities of future CAD/CAE systems addressing each physics faculty relevant to the component's functional fulfillment (e.g., mechanics, fluid-, thermodynamics).

e.g., sensitivity analysis as such is already common in the context of, e.g., *Computational Fluid Dynamics (CFD)* and performed in the early stage of CFD simulations in order to verify the simulation model and to determine the area of highest relevance for a targeted value (e.g., area of highest flow rate or lowest pressure losses in the volume). The computed results can be provided as envelop surfaces in 3D design space. Those results should not be considered as design geometries ready for detailing. Those envelopes only indicate design space segments of high importance for fulfilling a particular function, in this example, a function from the category *fluid transfer*.

As elaborated in Sect. 4.2.2 for mechanical load transfer, this sensitivity analysis might be derived from principle stresses or early-stage topology optimization results. For additional methods helpful to conduct this step, also consult [GTV18; Wib21; BBM22]. Once a sensitivity analysis has been conducted for each function to be realized in the same design space of the chosen component, those individually computed results should be superposed to a 3D sensitivity map of the design space. This way, areas of cross-functional conflicts in the design space of interest are revealed.

Step 2B: Monolythic embodyment:

Based on the 3D sensitivity map with identified conflict zones in the design space, *step 2B* strives towards an optimal compromise among the respective functions and their design space disputations. This step is focused on the component level and comprehended as the transition to the part level in the design workflow. This design task should be conducted from a monolithic perspective at the start to consider the functional integration potential of AM. The focus here is on the embodiment of the entire component as a one-piece-design (compare *one-piece-machine strategy* [EM13]), with, e.g., material placement according to load transfer intersections or material absence according to fluid/medium transfer intersections. Direct and feature-based design principles might be chosen here (e.g., a wrap of the envelope surfaces of the sensitivity map to start with).

Sub-concepts might be necessary to solve the previously identified conflict zones in the respective design space. Conventionally, those conflict zones do not occur due to disciplinary instead of interdisciplinary design approaches or are defused by repri-oritization of the respective functions, e.g., safety-relevant load transfer is highly prioritized over aerodynamic performance. Considering AM design potentials, a downgrading of a function might not be necessary. Applying geometric design potentials, varying materials, or material properties in the same part are a few exem-plary design principles that enable engineers to find compromises for design space disputations, renouncing general function downgrading and enhancing functional integration.

Subsequently, the partitioning of the component into individual parts needs to be considered, e.g., due to inevitable relative movements between parts, incompatible materials, ease of assembly, and ease of repair. The necessary part transitions should be set wisely, taking appropriate manufacturing process-specific restrictions into account. Considering their cost-intensive characteristics for large-scale production, AM technologies should be considered for complex and otherwise manufacturable parts. Depending on the geometric complexity level or the parametric extent of the design task, a suitable design principle, e.g., visual scripting, should be chosen for each part as preparation for the design detailing step (see Fig. 6.8 step 2B).

Step 3: Part Design:

Now, on the part level of the product structure, the detailed 3D part design is elabo-rated. This step focuses on a particular partition of the previous monolithic embod-iment and should be repeated for each partition of the blueprint. The partitions set and the sub-concepts developed by the design engineer can also be considered as a new design space definition and boundary conditions on the part level if a *generative design principle* is chosen further to determine the optimal exact geometry of the part. This workflow presupposes that CAE functionalities (e.g., topology optimization) are seamlessly integrated and provide native CAD data as design results to be detailed (compare Sect. 6.2.1).

Independently of the design principle of choice, e.g., direct or feature-based modeling, scripted, visually scripted, or algorithmically generated (compare 2.10), due to the sensitivity analysis and the embodiment design as a blueprint, design engineers do not necessarily rely on the 2D design of auxiliary geometry as starting point for their part design anymore. In the best case, a seamless transition between different design principles is enabled by the CAD system, allowing design engineers to switch to the most appropriate design principle for each particular task.

Additionally, functionalities towards the parametric design of intersections (e.g., flanges, mounting points, welding seams) should be implemented in the feature set parametrically. Due to the top-down approach of setting component partitions in the previous *step 2B*, partition changes on the component level should automatically lead to adaptions of the part-level design history and vice versa, enabling this way agile design adaptions and minimizing part collisions on component level.

According to the descriptions of the conventional design workflow in Sect. 2.1.2, also the envisioned design workflow continues with an iterative performance validation of the designed parts and components in *step 4* by *simulation and optimization* functionalities closely related to the conventional approach. The design workflow is finalized with the *export of production files and documentation data* in *step 5*. To carry out the envisioned approach, a multiphysics engineering software platform should be considered that seamlessly combines CAD and CAE functionalities. Additionally, partially implemented or upcoming design functionalities, such as fully integrated visual scripting in established engineering software platforms, will assist in applying complex AM-related design potentials.

Summarizing the detailed workflow described in Fig. 6.8, Fig. 6.10 illustrates only the main additional design steps in comparison to the conventional design workflow. Additionally, to improve comprehensibility, the interim results of the MDO vehicle segment concept elaborated in Sect. 4.2.2 are assigned to each additional step of the workflow.

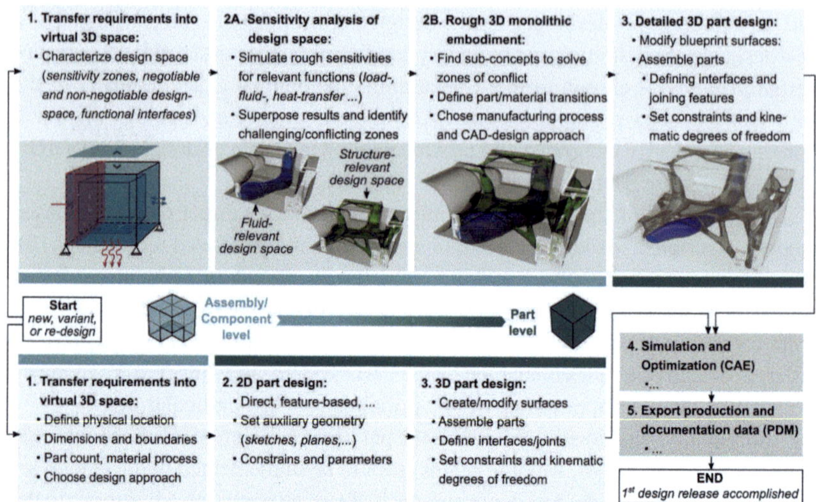

Fig. 6.10 Conventional parametric-associative cad workflow (bottom) opposed to proposed exemplary visionary DfAM inspired engineering design workflow (top) based on Fig. 6.8

6.3.3   Summary and Discussion

The major findings of this section can be summarized in the generalization of the approach elaborated for the conception of an MDO vehicle segment in Sect. 4.2.2. For this purpose, and to address the identified need for a design workflow that enables the utilization of DfAM potentials fully, a general procedure has been synthesized for the three exemplary physics disciplines load, fluid, and heat transfer (Fig. 6.6 and 6.7). One major objective of this approach is the identification of design space disputation zones between optima of different functions with high functional integration potential and finding conflict mitigation strategies considering DfAM design potentials accordingly (Fig. 6.6). Further, the approach has been translated into a step-by-step design workflow for engineers focusing on functional complex applications based on the characterization and sensitivity analysis of the design space. Subsequently, differences to the conventional part design workflow have been indicated (Fig. 6.8).

However, besides the elaborations in this work, the suggested MDO approach fit for DfAM still needs to be fully validated. On the one hand, this aspect is partially substantiated by missing functionalities in today's CAD and CAE tools (compare Sect. 6.2). Therefore, the suggested approach remains visionary and yet to be fully applicable. On the other hand, further research towards resilient sensitivity analysis methods for additional disciplines is necessary. Also, the translation of the suggested mitigation strategies into coupling functionalities of MDO software tools to manage multiple optimization operations simultaneously is a crucial next step to enable the applicability of the approach. Furthermore, many questions about the impact of multidisciplinarity-focused engineering design on today's disciplinary structured corporate development environment arise. e.g., do marginal gains in product performance justify increased corporate engineering complexity?

Nonetheless, concluding this section, the elaborated approach enables design engineers to develop engineering solutions with high functional integration and lightweight focus with the potential to increase the degree of efficiency of the product while simultaneously considering the design freedom of AM technologies purposefully. Further, partial aspects of the approach also apply beyond the AM context. Here, it is important to emphasize that trusting optimization results blindly might lead to severe errors. Thus, design engineers need to develop additional competencies regarding the evaluation of multidisciplinary optimized geometries. Alterations in engineering practice result demanding adaptions already in the training of upcoming engineers.

Conclusions and Future Work 7

In this chapter, the main findings of this thesis are summarized and discussed, and their added value for research and engineering practice is put into perspective. Lastly, recommendations for future research are derived from the limitations of this work.

7.1 Summary and Discussion

AM technologies are facing product-related challenges, such as meeting surface quality standards, homogeneous material properties, dimensional accuracy, and process-related improvement areas, e.g., scalability and economic efficiency. Nevertheless, design freedom and manufacturing flexibility assigned to AM technologies can extend the available solution space for design engineers, recalibrating the technical feasibility of complex designs. This work analyzed the potential impact of this increased solution space on product development systematically, considering challenges within the intersection between the three fields of research: Design for Additive Manufacturing (DfAM), Vehicle Conception (VC), and Design Methodology (DM). The research challenges of these fields identified and addressed in this thesis are summarized in the following:

Supplementary Information The online version contains supplementary material available at https://doi.org/10.1007/978-3-658-49677-7_7.

D. Fuchs, *Exploration of the Influence of Additive Manufacturing Design Potentials on Vehicle Conception*, AutoUni – Schriftenreihe 179,
https://doi.org/10.1007/978-3-658-49677-7_7

7.1.1　Research Challenges Addressed in this Thesis

Challenges in Design for Additive Manufacturing (DfAM): In the literature, DfAM methods and tools are theoretically described, considering the restrictions and potentials of AM technologies on a holistic level. However, identifying purposeful and profitable applications for the technology is still challenging, depending on the industry, particularly in the automotive context. If not profitable, use cases need to leverage additional, conventionally not achievable, product- or process-related benefits to justify the additive approach. State-of-the-art AM applications often consider design potentials one-sided only, not always focusing on benefiting the product's overall performance, thereby not utilizing AM design potentials holistically. Use cases on complete vehicle level have yet to consider AM design potentials systematically. Therefore, the impact of AM design potentials on vehicle conception had yet to be displayed exhaustively. Further, increased manufacturable design complexity in AM, often referred to as *complexity for free*, demands advanced engineering methods and tools, complicating development processes significantly. Particularly in the complete vehicle level context, a correlation between limitations in engineering methods and tools to enable multi-disciplinary optimized AM applications has been identified.

Challenges in Vehicle Conception (VC): Besides the exhaustive literature on the technical sections of the vehicle, little contemporary VC-specific literature addressing research deficits on the complete vehicle level can be found. Due to the complexity of VC as an engineering discipline and the deep integration into companies' core business, comprehensive research in this field is impeded. However, VC engineering practice is facing product-related challenges towards modularity and digitization in upcoming vehicle platforms, e.g., due to the transition of internal combustion engine to battery electric vehicle drive-train, resulting in increased energy storage costs, curb weight, and therefore altered crash behavior. Also, process-related hurdles, such as the increased functional scope lead to complex development processes that are being reoriented towards systems engineering and function-driven development. VC challenges arise also due to the multi-disciplinarity of engineering tasks in highly disciplinary organizational structures. However, elaborating on the optimal design solution as a compromise between the multitude of conflicting vehicle requirements strongly depends on the design freedom of manufacturing techniques. Thus, process- and product-related vehicle innovations enabled by additional design freedom have been elaborated in this thesis.

Challenges in Design Methodology (DM): Contemporary DM research is focused on combining established design methods (engineering design frameworks) and specialization of particular approaches (DfX), leading to evolutionary methodological progress. Particularly in combination with CAX tools, continuous advances in optimization techniques based on *Finite Element Method* (FEM) pave the way for new approaches towards multi-disciplinary optimization and multi-physics design in DM. In engineering practice, advances in DM are enabling progress in function-oriented system engineering development approaches. However, manual-labor intense design workflows, cost-cutting in PD, or decreasing corporate complexity through product variant reduction versus mass individualization trends in society are additional fields of tension in practice. Those circumstances demand advances in complexity management with innovative design methods, processes, tools, and manufacturing technologies. To dissociate from traditional engineering design patterns, propel the imagination of engineers, and enhance the capabilities of their tools, it becomes clear that a paradigm change in engineering design is necessary to systematically take advantage of design freedom and product complexity enabled by AM. Further, alterations of the hands-on engineering design process considering potentials and interferences between potentially opposed AM design principles are inevitable.

In this regard, this thesis's main objective is to explore the influence of additive manufacturing design potentials on vehicle conception.

7.1.2 Findings in the Context of the Research Questions

The findings in this work cover a broad spectrum in the intersection of the multitude of DfAM potentials and the wide range of technical aspects relevant to vehicle conception. The findings of this thesis are summarized in the following according to the three research questions defined in the introduction.

1st Research Question: To answer the question of how to apply DfAM potentials oriented on customer-relevant product performance, first, the primary vehicle requirement and property structure has been assessed (Chap. 3) and, second, vehicle-specific applications elaborated (Chap. 4). One significant finding here is that DfAM potentials are most beneficial for vehicle performance when applied to engineering trade-offs resulting from restrictions of conventional manufacturing technologies. Those trade-offs represent the best compromise design engineers could find for the respective vehicle concept, considering the multitude of requirements, conventional design techniques, and manufacturing methods. Those trade-offs occur mainly due to conflicting objectives in requirements and their derived functions during vehicle conception. Therefore, the basic idea here is to apply DfAM potentials to mitigate those conflicting objectives between vehicle properties. It has been shown that significant conventionally not feasible product improvements can be achieved.

To carry out this basic idea, conflicting objectives in the requirement and functional structure of the vehicle concept need to be identified and prioritized according to their customer relevance. Affected vehicle components should be determined, and their general additive feasibility should be assessed. Next, improvement solutions can be conceptualized with DfAM methods considering AM potentials, their interdependencies, and restrictions. Those main steps compile a validated approach for applying DfAM potentials oriented on customer-relevant product benefits (Sect. 6.1).

With the developed general approach for applying DfAM potentials oriented on customer-relevant product benefits, one possible answer to the first research question has been found. The various elaborated exemplary use cases further show potential DfAM-based vehicle innovations and present answers to specific engineering challenges occurring during their conception.

2nd Research Question: The question of which significant vehicle properties can be improved by a systematic and holistic application of DfAM potentials during vehicle conception has been elaborated, first, by identifying the key application areas for DfAM potentials on the complete vehicle level (Sect. 3.2). For this purpose, major vehicle characteristics to be optimized have been determined, and according to their correlation with DfAM potentials, four key application areas for DfAM with improvement potential on VC are categorized. As a result of these findings, DfAM potentials can either be applied to improve a vehicle's energy conversion efficiency (*performance*), *package and comfort, individualization*, or *safety* properties. A multitude of 22 exemplary sub-concepts have been derived from this application areas showcasing alternative approaches to VC based on AM design freedom (ESM App. A.4).

Second, specific applications have been elaborated to assess the general feasibility of the generated design principles, focusing on *mass individualization* and on *functional integration*. One significant finding is the developed approach to minimize variant-related engineering design complexity by realizing parametric geometry generators covering customer input, design, and simulation of mass individualized vehicle components (Sect. 4.1). Further, it has been shown that combining multi-physics optimization approaches and AM functional integration potential leads to alternative material- and design space-efficient solutions to vehicle conception (Sect. 4.2).

Third, to analyze the significance of DfAM-based vehicle property improvements, a technical up-scaling analysis leveraging DfAM-enabled lightweight design and alternative vehicle body design on complete vehicle level has been conducted (Chap. 5). Significant findings here are that a reduction of up to 6% of the curb weight resulting in up to 4 seconds lap-time savings of an exemplary sports car can be achieved by substantial topology optimization and additive manufacturing of metallic vehicle components (Sect. 5.1). Further, 11 alternative approaches to AM-based vehicle body design have been derived and assessed, illustrating the potential impact of DfAM-driven vehicle body design (Sect. 5.2).

The systematic elaboration of key DfAM application areas based on essential vehicle properties gives a holistic overview of potential answers to the second research question. Additionally, solution approaches to critical engineering challenges essential to these application areas have been analyzed in detail at part level. Lastly, exemplary vehicle property improvements on the part level have been leveraged to complete the vehicle level to illustrate and partially quantify their significance to VC.

3rd Research Question: This thesis also addresses how contemporary engineering approaches, methods, and tools must evolve to enable design engineers to leverage AM design potentials for vehicle conception holistically.

First and foremost, regardless of the multitude of AM design potentials, only if those potentials are applied with a focus on product performance or process efficiency a sustainable customer- or company-relevant benefit is achieved. To ensure this guideline is considered, a methodological approach for product-benefit-oriented application of AM potentials focused on conflicting development objectives in vehicle conception has been elaborated in this work (Sect. 6.1). The major finding here is that AM design potentials are most beneficial on the complete vehicle level when their application diminishes engineering trade-offs between conflicting objectives already in the early conception phase of PD. The suggested and tested approach enables full traceability of a DfAM product innovation to the respective conflicting objective to be diminished, thereby putting the additional product benefit in focus. This approach is considered supplementary to the PDP and does not replace it.

The specific applications elaborated in this work with a focus on AM-enabled mass individualization, multidisciplinary functional integration, and optimization reveal limitations of contemporary computer-aided engineering tools throughout the architectural levels of the system (Chap. 4). The elaborations in Sect. 6.2 summarize the necessary advances in CAD geometry representation, manipulation, optimization, and product structure, considering the front, back end, geometry kernel, and processing unit as system levels. Those advances include, e.g., fully parametric and associative modeling by visual scripting, agile implicit and explicit geometry representation methods, and function-oriented product structures. Due to the digital process characteristics of AM, only if design engineers are enabled to digitally define the product's complex geometry accurately by enhanced design, simulation, and optimization capabilities the purposeful production of the component is made possible.

However, besides the product-benefit-oriented approach and advanced CAD and CAE functionalities, conventional development methods need to adapt regarding their interdisciplinarity. As elaborated in Sect. 4.2, a significant finding to apply AM functional integration potentials is that today's strictly separated technical vehicle sections by disciplinary system boundaries must intertwine. Even though challenging from a corporate organizational and engineering point of view, by intertwining system boundaries, novel integrated design solutions can claim less material and design space than conventional approaches. For this purpose, significant findings of Sect. 6.3 are, first, an approach for multidisciplinary design space optimization and, second, the resulting necessary adaptions in the engineering design workflow.

The elaborated approaches to product-benefit-oriented application of DfAM potentials, multidisciplinary design space optimization, and the necessary advances to computer-aided engineering tools summarize how contemporary development approaches, methods, and tools need to evolve to leverage AM potentials holistically and formulate the answer to the third research question of this thesis.

7.1.3 Contributions to Research

The contributions to research in this thesis are summarized according to the identified challenges in the three relevant research fields at the beginning of this section.

Contributions to the research field Design for Additive Manufacturing (DfAM): The necessity to consider the multitude of AM design principles for a holistic utilization of AM potentials has been mentioned and supported by DfAM methods in the literature. By pointing out conflicts with the developed matrix of conflicting AM potentials, an additional tool to the semantic network of AM design potentials [Kum+18] can be considered for holistic applications counteracting one-sided approaches (Sect. 3.3). Additionally, the approach of leveraging AM design potentials systematically from part to complete vehicle level is considered novel. It represents a concept that might be transferable to further complex use cases, enabling quantified assessments of AM technology's impact (Sect. 5.1). Addressing the still challenging identification of resilient use cases for AM technologies in the automotive context, a systematic approach for product-benefit-oriented application of AM potentials considering their multitude and interdependencies has been formulated in this thesis (Sect. 6.1).

Contributions to the research field Vehicle Conception (VC): Regarding the low quantity of VC-specific literature on the complete vehicle level, this thesis contributes to research by summarizing VC fundamentals in English. Moreover, focus points for customer-oriented application of AM potentials in the automotive context have been identified (Sect. 3.2). Engineering design aspects focusing on variant-related complexity management have been indicated, pointing out methods for mass individualizing automotive components (Sect. 4.1). Novel design principles towards functionally integrated vehicle components, particularly the MDO-based vehicle design approach, contribute to contemporary research in the DfAM context (Sect. 4.2). Consequently, alternative DfAM-inspired structural vehicle body design approaches have been pointed out (Sect. 5.2).

Contributions to the research field Design Methodology (DM): The emphasis of the elaborations in this thesis on multidisciplinary, computationally optimized, or fully parametric design approaches complements conventional design methodology with additional complexity levels. With tangible use cases, those approaches are demonstrated in a hands-on manner (Chap. 4). Among those approaches, a procedure for implementing customer-individuality into geometric part design has been formulated (Sect. 4.1). The parametric and automated design validation aspect has also been introduced to the mass individualized automotive context. Moreover, methodological procedures towards functional-, shape- and variant-related design complexity have been developed in this work (Sect. 6.1), indicating their benefits and the limitations of contemporary engineering methods and tools (Sect. 6.2).

7.1.4 Contributions to Engineering Practice

Considering that AM potentials can be categorized into four complexities (shape, hierarchic, function, and material complexity), implementing AM potentials challenges engineering practice, especially during the conception and design phase of the part instead of during tool development to ensure manufacturability. This work contributes to reducing engineering design complexity in the DfAM context. Particularly the emphasis on parametrization and alternative design principles, e.g., visual scripting, in the developed design approaches enable innovative vehicle applications while minimizing engineering design effort (Sect. 4.1). A step-by-step approach to parametric mass individualized automotive components has been provided accordingly (Sect. 4.1). Several DfAM applications introduced in this work can assist design engineers during part-identification, particularly considering contradicting AM design principles (Sect. 3.3) and the suggested approach for product-benefit-oriented application of AM potentials (Sect. 6.1).

The findings of this thesis further contribute to engineering practice by providing a comprehensive model of CAD system architecture, enabling design engineers to comprehend correlations between CAD functionalities and their limitations (Sect. 6.2). In conclusion, the suggested adaptions to the hands-on CAD design workflow assist design engineers in coping with complex multiphysics design challenges by optimizing design space conflicts while taking DfAM design potentials into account purposefully (Sect. 6.3).

7.1.5 Limitations

The limitations of this work can be structured into findings- and approach-related limitations. Findings-related limitations have already been partially discussed in the respective chapter summaries. Therefore, the following subsections only discuss general findings-related limitations and focus on the approach-related limitations of this thesis.

Findings-related limitations: The findings in this work aim to address AM design potentials from a holistic perspective. However, AM technologies differ significantly in their technology readiness level and suitability for different application areas. This aspect includes, among others, discrepancies regarding the achievable part quality, choice of material and their properties, productivity, manufacturing costs, and pre-processing and post-processing efforts. The findings in this thesis are formulated from a general perspective and do not address those differences between AM technologies in particular. Therefore, when combined with unsuitable AM processes, the suggested use cases might lead to additional development and production challenges and thus end up in negative business cases. Further, considering the high-quality standards in the industry, the technology readiness level of AM technologies limits the feasibility of ambitious automotive applications in general, significantly challenging each area of AM application in VC suggested in this work.

Considering, on the one hand, the product diversity in the automotive industry and, on the other hand, the cost-intensive nature of AM technologies, the AM potentials for VC identified and developed in this work might initially lead to positive business cases for premium or performance automotive segments only. Further, the increased part complexity of AM applications leads to additional engineering design efforts during PDP. In contrast, product development efforts are steadily minimized due to the industry's competitive environment, leading to high cost pressure. Thus, indirect causal effects of increased part complexity, e.g., weight or design space savings, improved aesthetics, part count reduction, decreased assembly effort, and fewer storage costs, to name a few, need to be offset to justify increased AM-related engineering design effort during PDP.

The discrepancy between the mass individualization potentials assigned to AM technologies and the suitability of most AM technologies to manufacture only small lot sizes efficiently can be controversial. Mass individualized automotive one-off components might be realized by combining AM and parametric geometry generators. Nonetheless, depending on the target vehicle, producing those unique parts still sums up to large lot sizes, challenging the productivity of AM technologies further.

Approach-related limitations: This work aimed to assess the influence of AM design potentials on VC, implying a wide field of research within the intersection between AM, VC, and DM. Therefore, the research questions are posed from a holistic perspective. The aspiration to cover the multitude of AM technologies and related design potentials and the extensive technical complexity of automotive vehicles holistically limits the in-depth quantified assessment of the individual findings to some extent. For this reason, the general approach chosen for this thesis is based on identifying AM potentials on the complete vehicle level. However, the feasibility of those potentials are only assessed on a component-specific level. If reasonable, the findings have been subsequently leveraged to complete vehicle level as illustrated in Fig. 1.4. Even though this approach was chosen for complexity management purposes, it is less accurate than developing an entirely DfAM-focused vehicle concept for direct comparison with a conventional concept. Only in this way the full extent of AM design potentials and their causal effects, e.g., design space savings and part count reduction, are quantifiable. However, the complexity of a one-to-one accurate approach would exceed the limited scope of this PhD thesis.

Moreover, this approach has uncovered engineering design challenges on the application level that had not been addressed yet. Thereby, the field of tension is addressed, that part geometries can be additively manufactured, which engineers cannot design purposely with reasonable effort. Hence, besides assessing the influence of AM design potentials on VC, this thesis also focuses on necessary advancements in engineering design approaches, workflows, and tools. This aspect limits the feasibility and applicability of the suggested approaches, as these approaches focus on AM design potentials and less on the solutions space that conventional design approaches and tools enable.

Not all identified AM potential application areas for VC (performance, package and comfort, individualization, and safety) have been assessed to equal depth. The performance-focused aspect of AM-enabled lightweight design could be leveraged to the complete vehicle level in a quantified manner. Leveraging the remaining three AM application areas to complete vehicle level required to address significant design challenges in the context of variant- and function-related design complexity first. Subsequently, with further advancements in complex multi-physics optimization and advanced vehicle parametrization approaches, the holistic influence of the remaining application areas for AM in VC can be leveraged.

This thesis's general approach does not emphasize the validation of the derived approaches to cope with the revealed design challenges in the DfAM context. Therefore, these approaches should not be considered finalized methods ready for application. This limitation is partially due to further necessary advancements in engineering software tools limiting the tangible application of the respective approaches.

Lastly, a holistic assessment of AM technologies in the automotive context should include product- and process-related value propositions. Even though process-related aspects have been excluded from this work's scope, they still bear crucial value for the industry, e.g., improvements in time-to-product and time-to-market.

7.2   Outlook and Recommendations for Future Work

The content of this work presents many points of reference for future research. Considering AM technologies in general, future work should focus on improving the technological readiness level of each process with a focus on quality, consistency, and efficiency. This way, significant cost drivers can be counteracted, motivating additional research toward ambitious fields of applications. For now, a strong focus should be on realizing AM functional prototypes and after-sales applications. Those application areas are subject to series requirements yet only produced in low volumes. Future work should further assess the findings of this work, including use cases, application areas, and design approaches, regarding comprehensiveness, validity, and applicability to other engineering fields.

DfAM in general: DfAM can benefit from further advancements in DM addressing complexity management methods for function- and variant-related product complexity. This way, a foundation for a holistic application of AM design potentials is set. Also, general progress in multiobjective and multiphysics optimization methods is necessary in this context.

AM-enabled functional complexity, for instance, would profit from comprehensive research on the feasibility of functional elements (e.g., threads, tracks, bearings, bushings, seals) for each specific AM process. Subsequently, DfAM-inspired redesigns of conventional functional elements are necessary. Functional integration potential can be maximized further by progress in multiphysics optimization strategies that rely not only on prioritizing individual physical disciplines or functions. The first step here is defining the most efficient sensitivity analysis method for the remaining physical disciplines other than mechanics and fluid dynamics. Next, implementing algorithmic solution strategies to design space disputations between individual sensitivity results (local optima) into multiphysics software tools is considered necessary.

Addressing AM potentials towards hierarchical complexity, novel optimization approaches for purposeful design of porous mesoscopic structures (e.g., lattices, honeycombs, triply periodic minimal surfaces) are crucial. Those approaches should consider, e.g., optimal buckling stabilization of load-bearing shells, pressure loss reduction of flown-through structures, or heat transfer maximization while minimizing material investment. This way, not only monolithically optimized topologies but also bone-like structures can be purposefully designed to unveil additional, e.g., lightweight potential that is not feasible with conventional technologies. Closely related, additional advancements in the area of material complexity are worth pointing out. Approaches to tailor the material properties of single or multi-material parts on a voxel basis, e.g., by altering energy input in the part's cross-section, have still to be developed.

Considering AM-enabled shape complexity, alternative design strategies for complex parametric geometries are needed to ensure the stability and robustness of the geometry's representation. Here, automated calculation of compatible and incompatible parameter variations leading to valid closed surface design geometrics would significantly reduce the validation time of mass individualized parts. Particularly in this aspect, artificial intelligence algorithms might be disruptive.

Transferring DfAM potentials into VC engineering practice: The elaborations of this work are inspired by AM-related design freedom and manufacturing flexibility. The resulting optimizations or additions to engineering design approaches and tools are not necessarily only helpful in the DfAM context. Also, DfX on conventional manufacturing technologies might benefit from design approaches oriented on individualization, optimization, and functional integration. This aspect means that the use cases suggested in this work developed as a result of DfAM methods, might at least to some extent be combinable with conventional manufacturing technologies.

The reference points for future work mentioned above are fundamental to driving DfAM applications in the automotive context. Combined with enhanced CAD and CAE tools, disruptive approaches to vehicle design are possible. Besides AM lightweight potential, a next step here is to leverage also AM functional integration and mass customization to complete vehicle level, further quantifying the potential impact of DfAM design principles on VC. Additional assessment work that quantifies and offsets the beneficial causal effects of DfAM (e.g., weight or design space savings, part count reduction, decreased assembly, and logistic effort) from additional design and manufacturing costs is necessary to justify AM in the automotive context. This aspect leads to the process-related value propositions of AM (e.g., reduced time-to-product or time-to-market) yet to be addressed thoroughly, concluding the assessment of the impact of AM potentials for the automotive context.

Bibliography

[3Dn18]   3DNATIVES. *LSEV: Vollständig 3D-gedrucktes E-Auto für 2019?* Mar. 27, 2018. URL: https://www.3dnatives.com/de/lsev-3d-auto-270320181/#! (visited on 03/03/2024).

[Ada15]   ADAM, G. A. O. *Systematische Erarbeitung von Konstruktionsregeln für die additiven Fertigungsverfahren Lasersintern, Laserschmelzen und Fused Deposition Modeling.* Shaker Verlag, 2015.

[AHC15]   ANDREASEN, M. M., HANSEN, C. T., and CASH, P. *Conceptual design.* Springer, 2015.

[Alt17]   ALTAIR. *3i-PRINT—individualize, integrate, innovate.* Tech. rep. Nov. 1, 2017.

[AM97]   ANDREASEN, M. M. and MORTENSEN, N. H. "Basic Thinking Patterns and Working Methods for multiple DFX". In: *Proceedings of the 8th Symposium on Design for Manufacturing.* Ed. by MEERKAMM, H. Schnaittach, 1997.

[And+12]   ANDERL, R., EIGNER, M., SENDER, U., and STARK, R. *Smart Engineering: Interdisziplinäre Produktentstehung (acatech DISKUSSION).* Berlin Heidelberg: Springer, 2012.

[AR11]   ABELE, E. and REINHART, G. *Zukunft der Produktion: Herausforderungen, Forschungsfelder, Chancen.* Hanser, 2011.

[ASM17]   ASME. *ASME Standards: ASME Y14.46 Product Definition for Additive Manufacturing.* Tech. rep. American Society of Mechanical Engineers, 2017.

[Atl16]   ATLAS, N. *Honda 3D prints compact delivery EV.* Oct. 11, 2016. URL: https://newatlas.com/honda-kabuku-3d-printed-micro-commuter-delivery-vehicle/45852/ (visited on 03/04/2024).

[Awd+18]   AWD, M., TENKAMP, J., HIRTLER, M., SIDDIQUE, S., BAMBACH, M., and WALTHER, F. "Comparison of Microstructure and Mechanical Properties of Scalmalloy (registered) Produced by Selective Laser Melting and Laser Metal Deposition". In: *Materials* 11.1 (2018).

[AZ14]   ADAM, G. A. and ZIMMER, D. "Design for Additive Manufacturing—Element transitions and aggregated structures". In: *CIRP Journal of Manufacturing Science and Technology* 7.1 (2014), pp. 20–28.

[AZ15]   ADAM, G. A. O. and ZIMMER, D. "On design for additive manufacturing: evaluating geometrical limitations". In: *Rapid Prototyping Journal* 21.6 (2015), pp. 662–670.

© The Editor(s) (if applicable) and The Author(s) 2026

D. Fuchs, *Exploration of the Influence of Additive Manufacturing Design Potentials on Vehicle Conception*, AutoUni – Schriftenreihe 179,
https://doi.org/10.1007/978-3-658-49677-7

[Bal+12]   BALESDENT, M., BÉREND, N., DÉPINCÉ, P., and Chriette, A. "A survey of multidisciplinary design optimization methods in launch vehicle design". In: *Structural and Multidisciplinary Optimization* 45 (May 2012), pp. 619–642.

[Bar22]   BARTZ, R. "Eulersche Formoptimierung und automatisierte überführung topologieoptimierter Strukturbauteile inmodifizierbare Konstruktionsmodelle". PhD thesis. 2022.

[Bar+22]   BARTZ, R., FRANKE, T., FIEBIG, S., and VIETOR, T. "Density-based shape optimization of 3D structures with mean curvature constraints". In: *Structural and Multidisciplinary Optimization* 65 (2022), pp. 1–21.

[Bau03]   BAUER, S. "Design for X–Ansätze zur Definition und Strukturierung". In: *DFX 2003: Proceedings of the 14th Symposium on Design for X, Neukirchen/Erlangen, Germany, 13.-14.10. 2003*. 2003, pp. 1–8.

[Bau07]   BAUMBERGER, G. "Methoden zur kundenspezifischen Produktdefinition bei individualisierten Produkten". PhD thesis. Technischen Universität München, Jan. 2007.

[BBM21]   BIEDERMANN, M., BEUTLER, P., and MEBOLDT, M. "Automated design of additive manufactured flow components with consideration of overhang constraint". In: *Additive Manufacturing* 46 (2021), p. 102119.

[BBM22]   BIEDERMANN, M., BEUTLER, P., and MEBOLDT, M. "Routing multiple flow channels for additive manufactured parts using iterative cable simulation". In: *Additive Manufacturing* 56 (May 2022), p. 102891.

[BCP12]   BIN MAIDIN, S., CAMPBELL, I., and PEI, E. "Development of a design feature database to support design for additive manufacturing". In: *Assembly Automation* 32.3 (2012), pp. 235–244.

[Bec18]   BECKMANN, F. "3D-Druck für die Automobile Kleinserie". In: *Inside 3D-Printing*. 2018.

[Bel+18]   BELKADI, F., VIDAL, L. M., BERNARD, A., PEI, E., and SANFILIPPO, E. M. "Towards an Unified Additive Manufacturing Product-Process Model for Digital Chain Management Purpose". In: *28th CIRP Design Conference*. Vol. 70. 28th CIRP Design Conference 2018, 23–25 May 2018, Nantes, France. 2018, pp. 428–433.

[Ben22]   BENTLEY, M. *Ground Breaking 3D Printed Gold Process Used in Bentley Batur*. Bentley Motors Media. Dec. 16, 2022. URL: https://www.bentleymedia.com/en/newsitem/1405-ground-breaking-3d-printed-gold-process-used-in-bentley-batur (visited on 03/02/2024).

[Ber12]   BERMAN, B. "3-D printing: The new industrial revolution". In: *Business Horizons* 55.2 (2012), pp. 155–162.

[BF18]   BARTON, D. C. and FIELDHOUSE, J. D. *Automotive Chassis Engineering*. Cham: Springer International Publishing, 2018, pp. 215–254.

[BFR17]   BERMANO, A. H., FUNKHOUSER, T., and RUSINKIEWICZ, S. "State of the Art in Methods and Representations for Fabrication-Aware Design". In: *Computer Graphics Forum* 36.2 (2017), pp. 509–535.

[BG20]   BENDER, B. and GÖHLICH, D. *Dubbel Taschenbuch für den Maschinenbau 2: Anwendungen*. Jan. 2020.

[BGH05]   BECKER, R., GRZESIAK, A., and HENNING, A. "Rethink assembly design". In: *Assembly Automation* 25.4 (2005), pp. 262–266. eprint: https://doi.org/10.1108/01445150510626370.

[BHD18]   BERNHARD, M., HANSMEYER, M., and DILLENBURGER, B. "Volumetric modelling for 3D printed architecture." In: *Advances in Architectural Geometry*. 2018, pp. 392–415.

[BLR13]   BALDINGER, M., LEUTENECKER, B., and RIPPEL, M. "Desktop Manufacturing: Strategische Relevanz generative Fertigungsverfahren". In: *Industrie Management* 2. ISBN 9783955450182 (2013).

[BR09]   BOURELL, D. and ROSEN, D. *Roadmap for Additive Manufacturing—Identifying the Future of Freeform Processing*. Tech. rep. The University of Texas at Austin, Jan. 2009.

[Bra97]   BRAESS, H. "Das Automobil im Spannungsfeld zwischen Wunsch, Wissenschaft und Wirklichkeit. 2". In: *Stuttgarter Symposium Kraftfahrwesen und Verbrennungsmotoren*. 1997, pp. 786–799.

[Bri09]   BRILL, M. *Parametrische Konstruktion mit CATIA V5: Methoden und Strategien für den Fahrzeugbau*. 2. München: Hanser, 2009.

[BS13]   BRAESS, H. and SEIFFERT, U. *Vieweg Handbuch Kraftfahrzeugtechnik*. ATZ/MTZ-Fachbuch. Vieweg+Teubner Verlag, 2013.

[BS21]   BHATIA, A. and SEHGAL, A. K. "Additive manufacturing materials, methods and applications: A review". In: *Materials Today: Proceedings* (2021).

[BSZ20]   BRIARD, T., SEGONDS, F., and ZAMARIOLA, N. "G-DfAM: a methodological proposal of generative design for additive manufacturing in the automotive industry". In: *International Journal on Interactive Design and Manufacturing (IJIDeM)* 14.3 (2020), pp. 875–886.

[Bug18]   BUGATTI, E. G. *Weltpremiere: Bremssattel aus dem 3D-Drucker*. Jan. 22, 2018. URL: https://newsroom.bugatti.com/de/pressemeldungen/weltpremiere-bremssattel-aus-dem-3d-drucker (visited on 03/02/2024).

[Car+21] CARROLL, S., LYNE, J., OTREBA, M., and MCKENNA, T. "Integrating Computational Design into Structural Engineering Workflows to enhance Design Automation". In: *CitA BIM Gathering 2021*. Sept. 2021.

[Cha+17] CHAMBON, P., CURRAN, S., HUFF, S., LOVE, L., POST, B., WAGNER, R., JACKSON, R., and GREEN, J. "Development of a range-extended electric vehicle powertrain for an integrated energy systems research printed utility vehicle". In: *Applied Energy* 191 (2017), pp. 99–110.

[Che+17] CHEN, L., HE, Y., YANG, Y., NIU, S., and REN, H. "The research status and development trend of additive manufacturing technology". In: *The International Journal of Advanced Manufacturing Technology* Ausgabe 9–12 (2017).

[Chr+15] CHRISTIANSEN, A. N., BÆRENTZEN, J. A., NOBEL-JØRGENSEN, M., AAGE, N., and SIGMUND, O. "Combined shape and topology optimization of 3D structures". In: *Computers & Graphics* 46 (2015). Shape Modeling International 2014, pp. 25–35.

[CKZ14]   COHRS, M., KLIMKE, S., and ZACHMANN, G. "Streamlining Function-oriented Development by Consistent Integration of Automotive Function Architectures with CAD Models". In: *Computer-Aided Design and Applications* 11 (Feb. 2014), pp. 399–410.

[Cla17a] CLARKE, C. *Deutsche Bahn Get On-Board Local Motor's Self-Driving Olli in Berlin*. Jan. 16, 2017. URL: https://3dprintingindustry.com/news/deutsche-bahn-get-board-local-motors-self-driving-olli-berlin-103300/ (visited on 03/05/2024).

[Cla17b] CLARKE, C. *Local Motors 3D Prints First Olli, Autonomous Transit Vehicle in Knoxville*. May 11, 2017. URL: https://3dprintingindustry.com/news/local-motors-3d-prints-first-olli-autonomous-transit-vehicle-knoxville-113041/ (visited on 03/05/2024).

[Con+14] CONNER, B. P., MANOGHARAN, G. P., MARTOF, A. N., RODOMSKY, L. M., RODOMSKY, C. M., JORDAN, D. C., and LIMPEROS, J. W. "Making sense of 3-D printing: Creating a map of additive manufacturing products and services". In: *Additive Manufacturing* 1–4 (2014). Inaugural Issue, pp. 64–76.

[CSI17] CSI. *Press release: Youngtimer trifft Zukunftstechnologie: VW Caddy erhält funktionsintegrierten 3D-Druck Vorderwagen—Individualisieren, integrieren, Innovationen treiben: Partnerprojekt 3i-PRINT zeigt, was mit industriellem 3D-Druck möglich ist*. Tech. rep. CSI Entwicklungstechnik GmbH, 2017.

[CSI18] CSI. "Via 3D-Druck zum KFZ-Rahmen: Additive Fertigung schreitet voran". In: *Welt der Fertigung—Das Magazin für Praktiker und Entscheider* 04 (2018), p. 72.

[Die15] DIEHL, S. *Peugeot Fractal Concept (IAA 2015): Sitzprobe—Der fährt auf der Tonspur*. Sept. 3, 2015. URL: https://www.autobild.de/artikel/peugeot-fractal-concept-iaa-2015-sitzprobe-6013257.html (visited on 03/04/2024).

[DIN03] DIN. *Deutsche Norm: DIN 8580 Fertigungsverfahren—Begriffe, Einleitung*. Tech. rep. Deutsches Institut für Normung, 2003.

[DIN20] DIN. *Deutsche Norm: DIN 8580 Fertigungsverfahren—Begriffe, Einleitung (Entwurf)*. Tech. rep. Deutsches Institut für Normung, 2020.

[DM+20] DELGADO-MACIEL, J., CORTÉS-ROBLES, G., SÁNCHEZ-RAMÍREZ, C., GARCÍA-ALCARAZ, J., and MÉNDEZ-CONTRERAS, J. M. "The evaluation of conceptual design through dynamic simulation: A proposal based on TRIZ and system Dynamics". In: *Computers & Industrial Engineering* 149 (2020), p. 106785.

[DPL20] DALPADULO, E., PINI, F., and LEALI, F. "Integrated CAD platform approach for Design for Additive Manufacturing of high performance automotive components". In: *International Journal on Interactive Design and Manufacturing (IJIDeM)* (Aug. 2020).

[DPL21] DALPADULO, E., PINI, F., and LEALI, F. "Assessment of Computer-Aided Design Tools for Topology Optimization of Additively Manufactured Automotive Components". In: *Applied Sciences* 11.22 (2021).

[DVH12] DOUBROVSKI, E. L., VERLINDEN, J. C., and HORVATH, I. "First Steps Towards Collaboratively Edited Design for Additive Manufacturing Knowledge". In: *Solid Freeform Fabrication Symposium*. Austin, TX, 2012, pp. 891–901.

[EDA15] EDAG. *Bionische Fahrzeugkonzepte: Blaupausen aus der Natur für die Automobilindustrie*. Aug. 6, 2015. URL: https://www.edag.com/de/edag-group/presse/pressemeldung/bionische-fahrzeugkonzepte-blaupausen-aus-der-natur-fuer-die-automobilindustrie (visited on 03/03/2024).

[Ehr+14] EHRLENSPIEL, K., KIEWERT, A., LINDEMANN, U., and MÖRTL, M. *Kostengünstig Entwickeln und Konstruieren: Kostenmanagement bei der integrierten Produktentwicklung*. 7. Berlin Heidelberg: Springer Vieweg, 2014.

[Eib16] EIBISCH, H. "Additive Manufacturing of Structural Parts, Technological Challenges from an Automotive Manufacturer's Point of View". In: *Internationale Luftfahrtausstellung Berlin*. 2016.

[Eil99]   EILETZ, R. "Zielkonfliktmanagement bei der Entwicklung komplexer Produkte—am Bsp. PKW-Entwicklung". PhD thesis. Technische Universität München, 1999.

[EM13]   EHRLENSPIEL, K. and MEERKAMM, H. *Integrierte Produktentwicklung: Denkabläufe, Methodeneinsatz, Zusammenarbeit.* 5. München Wien: Hanser, 2013.

[Emm+11]   EMMELMANN, C., SANDER, P., KRANZ, J., and WYCISK, E. "Laser Additive Manufacturing and Bionics: Redefining Lightweight Design". In: *Physics Procedia* 12, Part A (2011), pp. 364–368.

[Eva22]   EVANS, S. *2023 Czinger 21C Prototype First Ride: One Wild Creation.* May 31, 2022. URL: https://www.motortrend.com/reviews/2023-czinger-21c-3d-printed-hypercar-prototype-first-ride-review/ (visited on 03/05/2024).

[Fer+18]   FERA, M, MACCHIAROLI, R, FRUGGIERO, F, and LAMBIASE, A. "A new perspective for production process analysis using additive manufacturing—complexity vs production volume". In: *The International Journal of Advanced Manufacturing Technology* 95.1 (2018), pp. 673–685.

[FFK19]   FIEGL, T., FRANKE, M., and KÖRNER, C. "Impact of build envelope on the properties of additive manufactured parts from AlSi10Mg". In: *Optics & Laser Technology* 111 (2019), pp. 51–57.

[FG13]   FELDHUSEN, J. and GROTE, K. *Pahl/Beitz Konstruktionslehre: Methoden und Anwendung erfolgreicher Produktentwicklung.* Springer Berlin Heidelberg, 2013.

[FHK20]   FUCHS, D., HARTMANN, I., and KUMKE, M. *Verfahren zur Erstellung von Realisierungskonzepten für Bauteilgeometrien unter Berücksichtigung verschiedener physikalischer Energieflüsse.* Tech. rep. DE 10 2020 116 463 A1 2021.12.23. Deutsches Patent- und Markenamt, 2020.

[Fie16]   FIEBIG, S. "Form-und Topologieoptimierung mittels Evolutionärer Algorithmen und heuristischer Strategien". ISBN 978-3-8325-4367-9. PhD thesis. 2016.

[Foe09]   FOERSTER, A. "Lebenstile als Instrument zur Segmentierung von Markt und Maken—ein Fallbeispiel in einem deutsch-französischen Unternehmen der Automobilbranche". PhD thesis. Technische Universität Kaiserslautern, 2009.

[Fra09a]   FRANKE, H.-J. "Grundlagen der Produktentwicklung und Konstruktion". Skript zur Vorlesung. 2009.

[Fra09b]   FRANKE, H. "Zielkonflikte bei der Karosserieentwicklung: Strategien und Methoden zu ihrer Beherrschung". In: *Vortrag Tagung: Strategische Fragen des Karosseriebaus.* 2009.

[Fra18]   FRANKE, T. "Fertigungsgerechte Bauteilgestaltung in der Topologieoptimierung auf Grundlage einer integrierten Gießsimulation". PhD thesis. 2018.

[Fri13]   FRIEDRICH, H. *Leichtbau in der Fahrzeugtechnik.* ATZ/MTZ-Fachbuch. Springer Fachmedien Wiesbaden, 2013.

[Fuc+20]   FUCHS, D., KUSCHMITZ, S., KÜHLKE, K., and VIETOR, T. "Identifikation von Zielkonflikten bei der Anwendung von Potenzialen additiver Fertigungsverfahren". In: *Konstruktion für die Additive Fertigung 2019.* Ed. by Lachmayer, R., Rettschlag, K., and Kaierle, S. Berlin, Heidelberg: Springer Berlin Heidelberg, 2020, pp. 223–244.

[Fuc+22]   FUCHS, D., BARTZ, R., KUSCHMITZ, S., and VIETOR, T. "Necessary Advances in Computer-Aided Design to Leverage on Additive Manufacturing Design Freedom". In: *International Journal on Interactive Design and Manufacturing* (2022).

[Fun15]   FUNG, D. *Local Motors LM3D Swim launched, 3D printed car on sale 2016.* Nov. 7, 2015. URL: https://www.drive.com.au/news/local-motors-lm3d-swim-3d-printed-car-launched-on-sale-2016/ (visited on 03/05/2024).

[Gao+15]   GAO, W., ZHANG, Y., RAMANUJAN, D., RAMANI, K., CHEN, Y., WILLIAMS, C. B., WANG, C. C., SHIN, Y. C., ZHANG, S., and ZAVATTIERI, P. D. "The status, challenges, and future of additive manufacturing in engineering". In: *Computer-Aided Design* 69 (2015), pp. 65–89.

[Geb16]   GEBHARDT, A. *Additive Fertigungsverfahren: Additive Manufacturing und 3D-Drucken für Prototyping—Tooling—Produktion.* Carl Hanser Verlag GmbH & Company KG, 2016.

[GH16]   GEBHARDT, A. and HÖTTER, J. *Additive Manufacturing: 3D Printing for Prototyping and Manufacturing.* Carl Hanser Verlag GmbH & Company KG, 2016.

[Gib+21]   GIBSON, I., ROSEN, D., STUCKER, B., and KHORASANI, M. *Additive Manufacturing Technologies.* Third Edition. Springer Nature Switzerland AG, 2021.

[GK10]   GUSIG, L. and KRUSE, A. *Fahrzeugentwicklung im Automobilbau: aktuelle Werkzeuge für den Praxiseinsatz ; mit 28 Tabellen und 55 übungsfragen.* Hanser, 2010.

[GK19]   GUNDLACH, I. and KONIGORSKI, U. "Modellbasierte Online-Trajektorienplanung für zeitoptimale Rennlinien". In: *AT—Automatisierungstechnik* 67.9 (2019), pp. 799–813.

[GKT16]   GEBHARDT, A., KESSLER, J., and THURN, L. *3D-Drucken: Grundlagen und Anwendungen des Additive Manufacturing (AM).* Carl Hanser Verlag GmbH & Company KG, 2016.

[GKT19]   GEBHARDT, A., Kessler, J., and Thurn, L. *3D Printing.* Second Edition. Hanser, 2019.

[GL15]   GEMBARSKI, P. and LACHMAYER, R. "Degrees of Customization and Sales Support Systems—Enablers to Sustainablility in Mass Customization". In: *International Conference on Engineering Design, ICED 15.* Milano, Italy, July 2015.

[Glu18]   GLUCKER, J. *Hackrod Uses 3D Printing and VR to Bring Beatufil La Bandita to Life.* Mar. 27, 2018. URL: https://www.motorauthority.com/news/1115933_could-lincoln-bring-back-suicide-doors-for-a-future-continental (visited on 03/05/2024).

[Gri+19]   GRIGOLATO, L., ROSSO, S., MENEGHELLO, R., CONCHERI, G., and SAVIO, G. "Heterogeneous objects representation for Additive Manufacturing: a review". In: *Instant Journal of Mechanical Engineering* (Dec. 2019), pp. 14–23.

[GRS10]   GIBSON, I., ROSEN, D., and STUCKER, B. *Additive Manufacturing Technologies: 3D Printing, Rapid Prototyping, and Direct Digital Manufacturing.* Springer New York, 2010.

[GRS16]   GIBSON, I., ROSEN, D., and STUCKER, B. *Additive Manufacturing Technologies: 3D Printing, Rapid Prototyping, and Direct Digital Manufacturing.* Springer New York, 2016.

[GTV18]   GEBHARDT, P., TÜRCK, E., and VIETOR, T. "A Lean Method for Local Patch Reinforcement Using Principal Stress Lines". In: *Advances in Structural and Multidisciplinary Optimization—Proceedings of the 12th World Congress of Structural and Multidisciplinary Optimization (WCSMO12)* (2018), pp. 789–798.

[Hak13]   HAKEN, K. *Grundlagen der Kraftfahrzeugtechnik: Fahrzeugtechnik*. Hanser, 2013.

[Hak18]   HAKEN, K. *Grundlagen KFZ-Technik, 5. a.* Carl Hanser GmbH, 2018.

[HCD03]   HAGUE, R., CAMPBELL, I., and DICKENS, P. "Implications on design of rapid manufacturing". In: *Proceedings of the Institution of Mechanical Engineers, Part C: Journal of Mechanical Engineering Science* 217.1 (2003), pp. 25–30.

[HEG07]   HEIßING, B., Ersoy, M., and Gies, S. *Fahrwerkhandbuch*. Springer, 2007.

[Hel+18]   HELD, M., WEHNER, D., HÄMMERL, R., DANGELMAIER, M., BRIEM, A.-K., REIFF, C., and WULLE, F. "Personalization in the Automotive and Building Sector—Research Program of the High-performance Center »Mass Personalization« in Stuttgart". In: *8th International Conference on Mass Customization and Personalization—Community of Europe*. Sept. 2018.

[Hen13]   HENKER, E. *Fahrwerktechnik: Grundlagen, Bauelemente, Auslegung*. Springer-Verlag, 2013.

[HM11]   HENNING, F. and MOELLER, E. *Handbuch Leichtbau: Methoden, Werkstoffe, Fertigung*. Hanser, 2011.

[HMS04]   HAGUE, R., MANSOUR, S., and SALEH, N. "Material and design considerations for rapid manufacturing". In: *International Journal of Production Research* 42.22 (2004), pp. 4691–4708.

[Hof19]   HOFMANN, R. G. "Integrationsgehäuse für elektrischen Achsantrieb, a exhibition prototype from the company Fa. Robert Hofman GmbH". In: *Formnext*. 2019.

[HP18]   HP, D. C. *With HP Metal Jet, Volkswagen Accelerates Manufaturing*. Oct. 3, 2018. URL: https://reinvent.hp.com/us-en-3dprint-volkswagen-metaljet (visited on 03/02/2024).

[HSPI05]   HUANG, G. Q., SIMPSON, T. W., and PINE II, B. J. "The power of product platforms in mass customisation". In: *International Journal of Mass Customisation* 1.1 (2005), pp. 1–13.

[Hua+15]   HUANG, Y., LEU, C. L., MAZUMDER, J., and DONMEZ, A. "Additive Manufacturing: Current State, Future Potential, Gaps and Needs, and Recommendations". In: *Journal of Manufacturing Science and Engineering* (2015).

[Jae+20]   JAESCHKE, N., LINGNER, M., GOUSHEGIR, S. M., SENKEL, Y., and BECKMANN, F. *Bionischer LAM-Stahlleichtbau für den Automobilbau—BioLAS*. Research rep. 320. Fraunhofer-Einrichtung für Additive Produktionstechnologien IAPT, 2020.

[Kab+17]   KABASAKAL, I., KESKIN, F. D., VENTURA, K., and SOYUER, H. "From mass customization to product personalization in automotive industry: potentials of industry 4.0". In: *Journal of Management Marketing and Logistics* 4.3 (2017), pp. 244–250.

[Kal+22]   KALMBACH, R., STRICKER, K., CORREA, P., and FOUCAR, D. *Automotive and Mobility M&A: How Companies Tune Up Their M&A Engines*. Tech. rep. Bain & Company, 2022.

[KBF05]   KASIK, D. J., Buxton, W., and FERGUSON, D. R. "Ten CAD challenges". In: *IEEE Computer Graphics and Applications* 25.2 (2005), pp. 81–92.

[KBP07]   KOREN, Y., BARHAK, J., and PASEK, Z. *Method and Apparatus for Reconfigurable Vehicle Interiordesign and Business Transaction*. Tech. rep. US20070156540A1. United States Patent and Trademark Office, 2007.

[Key+12]   KEYES, D., WOODWARD, A., GROPP, W., BELL, I., BROWN, L., CLO, A., CONNORS, K., CONSTANTINESCU, E., ESTEP, D., EVANS, K., FARHAT, C., HAMMOND, P., HANSEN, G., HILL, S., ISAAC, T., KAUSHIK, D., KAXIRAS, E., KONIGES, A., LEE, L., and Wohlmuth, B. *Multiphysics Simulations: Challenges and Opportunities*. Tech. rep. Jan. 2012.

[KF21]   KÜHLKE, K. and FUCHS, D. *Haltevorrichtung für ein Fahrzeug, Verfahren zur Herstellung einer Haltevorrichtung sowie Fahrzeug*. Tech. rep. DE 10 2020 208 323 A1. Deutsches Patent- und Markenamt, 2021.

[KFV23]   KUSCHMITZ, S., FUCHS, D., and VIETOR, T. "Customer Benefit Oriented Approach on the Application of Additive Manufacturing Potentials Based on Product Property Trade-Off's". In: *Innovative Product Development by Additive Manufacturing 2021*. Ed. by LACHMAYER, R., BODE, B., and KAIERLE, S. Cham: Springer International Publishing, 2023, pp. 305–317.

[KG18]   KRAUSE, D. and GEBHARDT, N. *Methodische Entwicklung modularer Produktfamilien: Hohe Produktvielfalt beherrschbar entwickeln*. Springer Berlin Heidelberg, 2018.

[KG19]   KLEIN, B. and GÄNSICKE, T. *Leichtbau-Konstruktion: Dimensionierung, Strukturen, Werkstoffe und Gestaltung*. Springer Vieweg. in Springer Fachmedien Wiesbaden GmbH, 2019.

[Kor+15]   KOREN, Y., SHPITALNI, M., GU, P., and HU, S. "Product Design for Mass-Individualization". In: *CIRP 25th Design Conference Innovative Product Creation*. Vol. 36. CIRP 25th Design Conference Innovative Product Creation. 2015, pp. 64–71.

[Kor21]   KOREN, Y. "The Local Factory of the Future for Producing Individualized Products." In: (Nov. 2021).

[Kos+20]   KOSSIAKOFF, A., BIEMER, S., SEYMOUR, S., and FLANIGAN, D. *Systems Engineering Principles and Practice*. Wiley Series in Systems Engineering and Management. Wiley, 2020.

[KS14]   KÜÇÜKAY, F. and SEEKAMP, C. "Manuskript zur Vorlesung Fahrzeugkonstruktion 1, Fahrzeugantriebe, Kupplungen, Getriebe, Hybridstrukturen, Allradtechnik". 2014.

[Kuc12]   KUCHENBUCH, K. *Methodik Zur Identifikation und Zum Entwurf Packageoptimierter Elektrofahrzeuge*. Autouni—Schriftenreihe. Logos Verlag Berlin, 2012.

[Kum18]   KUMKE, M. "Methodisches Konstruieren von additiv gefertigten Bauteilen". PhD thesis. Technische Universität Braunschweig, 2018.

[Kum+18]   KUMKE, M., WATSCHKE, H., HARTOGH, P, BAVENDIEK, A., and VIETOR, T. "Methods and tools for identifying and leveraging additive manufacturing design potentials". In: *International Journal on Interactive Design and Manufacturing* (2018).

[Kum+21]   KUMKE, M., HARTMANN, I., FUCHS, D., DEMIRCAN, F., and WEHMANN, Y. *Modulares System und Verfahren zum individuellen Konfigurieren eines Kraftfahrzeugs*. Tech. rep. DE 10 2019 211 824 A1. Volkswagen AG. Deutsches Patent- und Markenamt, 2021.

[Kus23]   KUSCHMITZ, S. "Systematische Wissensvermittlung für die additive Fertigung am Beispiel von Konstruktionsprinzipien für die Materialextrusion". PhD thesis. Technische Universität Braunschweig, Institut für Konstruktionstechnik, 2023.

[KWV16]   KUMKE, M., Watschke, H., and Vietor, T. "A new methodological framework for design for additive manufacturing". In: *Virtual and Physical Prototyping* 11.1 (2016), pp. 3–19.

[Lea+14] LEARY, M., MERLI, L., TORTI, F., MAZUR, M., and Brandt, M. "Optimal topology for additive manufacture: A method for enabling additive manufacture of support-free optimal structures". In: *Materials & Design* 63 (2014), pp. 678–690.

[Li+18]   LI, J., NIE, Y., ZHANG, X., WANG, K., TONG, S., and EYNARD, B. "A Framework Method of User-participation Configuration Design for Complex Products". In: *28th CIRP Design Conference 2018*. Vol. 70. 28th CIRP Design Conference 2018, 23–25 May 2018, Nantes, France. 2018, pp. 451–456.

[Li+21]   LI, S., XIN, Y., YU, Y., and WANG, Y. "Design for additive manufacturing from a force-flow perspective". In: *Materials & Design* 204 (2021), p. 109664.

[Lin09]   LINDEMANN, U. *Methodische Entwicklung technischer Produkte: Methoden flexibel und situationsgerecht anwenden.* VDI-Buch. Springer Berlin Heidelberg, 2009.

[Lin16]   LINDEMANN, U. *Handbuch Produktentwicklung.* Carl Hanser Verlag GmbH & Company KG, 2016.

[LK21]   LI, G. and KOHLWES, P. *Produktivitätssteigerung und Kostensenkung der laseradditiven Fertigung für den Automobilbau—ProKos.* Tech. rep. 358. Fraunhofer-Einrichtung für Additive Produktionstechnologien IAPT, 2021.

[LL20]   LACHMAYER, R. and LIPPERT, R. *Entwicklungsmethodik für die Additive Fertigung.* Jan. 2020.

[LLK20]   LACHMAYER, R., LIPPERT, R. B., and KAIERLE, S. "Konstruktion für die Additive Fertigung 2018". In: Springer Berlin Heidelberg, 2020.

[LRZ06]   LINDEMANN, U., REICHWALD, R., and ZÄH, M. *Individualisierte Produkte—Komplexität beherrschen in Entwicklung und Produktion.* VDI-Buch. Springer Berlin Heidelberg, 2006.

[Lu+18]   LU, W., VIETOR, T., BLUMRICH, R., and WIEDEMANN, J. "Innovative electric vehicle concepts with optimized acoustic performance". In: *18. Internationales Stuttgarter Symposium.* Ed. by BARGENDE, M., REUSS, H.-C., and WIEDEMANN, J. Wiesbaden: Springer Fachmedien Wiesbaden, 2018, pp. 1305–1320.

[Lut23]   LUTZ, A. "Methodische Werkstoff- und Prozessentwicklung für die additive Serienproduktion von automobilen Strukturkomponenten". PhD thesis. Technischen Universität Hamburg, 2023.

[Mal19]   MALAN, A. *Fiat's Centoventi concept lets customer customize car.* Mar. 6, 2019. URL: https://europe.autonews.com/geneva-auto-show/fiats-centoventi-concept-lets-customer-customize-car (visited on 03/04/2024).

[Mas43]   MASLOW, A. "A Theory of Human Motivation". In: *Psychological Review* 50 (1943).

[Mat23]   MATHUR, P. "Creative Impact of an Event-driven Visual Scripting Tool". In: *CAADRIA Conference 2023.* Mar. 2023.

[MC05]   MACNEILL, S. and CHANARON, J.-J. "Trends and drivers of change in the European automotive industry:(I) mapping the current situation". In: *International journal of automotive technology and management* 5.1 (2005), pp. 83–106.

[Mei99]   MEINERS, W. "Direktes selektives Laser-Sintern einkomponentiger metallischer Werkstoffe". PhD thesis. RWTH-Aachen University, 1999.

[Mit+14] MITAL, A., DESAI, A., SUBRAMANIAN, A., and MITAL, A. *Product Development: A Structured Approach to Consumer Product Development, Design, and Manufacture*. Elsevier Science, 2014.

[Mor17] MORITZ, M. *3D-Druck haucht VW Caddy 1 wieder neues Leben ein*. Ed. by 3Dnatives. Aug. 31, 2017. URL: https://www.3dnatives.com/de/vw-caddy310820171/ (visited on 04/05/2024).

[MPW18] MARTIN, J. C., POLEWARCZYK, J. M., and WOLCOTT, P. J. *Eine Kraftfahrzeugstrukturkomponente und ein Herstellungsverfahren*. Tech. rep. DE102018104723. Deutsches Patent- und Markenamt, 2018.

[MWG09] MACEY, S., WARDLE, G., and GILLES, R. *H-point: The Fundamentals of Car Design & Packaging*. Art Center College of Design, 2009.

[MWJ17] MANI, M., WITHERELL, P., and JEE, H. "Design Rules for Additive Manufacturing: A Categorization". In: International Design Engineering Technical Conferences and Computers and Information in Engineering Conference Volume 1: 37th Computers and Information in Engineering Conference (Aug. 2017). V001T02A035.

[NA19] NAJIAN ASL, R. "Shape optimization and sensitivity analysis of fluids, structures, and their interaction using Vertex Morphing parametrization". PhD thesis. Technische Universität München, 2019.

[Nae18] NAEFE, P. *Konstruktionsmethodik: Kurz und bündig*. essentials. Springer Fachmedien Wiesbaden, 2018.

[Neh] NEHUIS, F. "Expert interview: Function- and part-oriented vehicle architecture". 14. February 2017.

[Neh14] NEHUIS, F. "Methodische Unterstützung bei der Ermittlung von Anforderungen in der Produktentwicklung". PhD thesis. 2014.

[Ngo+18] NGO, T. D., Kashani, A., Imbalzano, G., Nguyen, K. T., and Hui, D. "Additive manufacturing (3D printing): A review of materials, methods, applications and challenges". In: *Composites Part B: Engineering* 143 (2018), pp. 172–196.

[NL20] NAEFE, P. and LUDERICH, J. *Konstruktionsmethodik für die Praxis: Aktuelle Verfahren in der Produktentwicklung*. Jan. 2020.

[Orl17] ORLOFF, M. *ABC-TRIZ: Introduction to Creative Design Thinking with Modern TRIZ Modeling*. Springer International Publishing, 2017.

[ORN15] ORNL. *3-D printed Shelby Cobra highlights ORNL R&D at Detroit Auto Show*. Jan. 12, 2015. URL: https://www.ornl.gov/news/3-d-printed-shelby-cobra-highlights-ornl-rd-detroit-auto-show (visited on 03/04/2024).

[Ort+21] ORTMANN, C., SPERBER, J., SCHNEIDER, D., LINK, S., and SCHUMACHER, A. "Crashworthiness design of cross-sections with the Graph and Heuristic based Topology Optimization incorporating competing designs". In: *Structural and Multidisciplinary Optimization* 64 (2021), pp. 1063–1077.

[Pah+07] PAHL, G., BEITZ, W., FELDHUSEN, J., and GROTE, K. *Engineering Design: A Systematic Approach*. Solid mechanics and its applications. Springer London, 2007.

[PBD15] PREM, A., BASTIEN, C., and DICKISON, M. "Multidisciplinary Design Optimisation Strategies for Lightweight Vehicle Structures". In: *10th European LS-DYNA Conference 2015, Würzburg, Germany*. 2015.

[Peu15]   PEUGEOT. *FRACTAL: das PEUGEOT i-Cockpit akustisch verstärkt*. Sept. 4, 2015. URL: https://www.media.stellantis.com/de-de/peugeot/press/fractal-das-peugeot-i-cockpit-akustisch-verstarkt-1632576742-1441317600 (visited on 03/04/2024).

[PG98]   PHAM, D. and GAULT, R. "A comparison of rapid prototyping technologies". In: *International Journal of Machine Tools and Manufacture* 38.10–11 (1998), pp. 1257–1287.

[PNR18]   PRAKASH, K. S., NANCHARAIH, T., and RAO, V. S. "Additive Manufacturing Techniques in Manufacturing -An Overview". In: *Materials Today: Proceedings* 5.2, Part 1 (2018). 7th International Conference of Materials Processing and Characterization, March 17–19, 2017, pp. 3873–3882.

[Pon+14]   PONCHE, R., KERBRAT, O., MOGNOL, P., and HASCOET, J.-Y. "A novel methodology of design for Additive Manufacturing applied to Additive Laser Manufacturing process". In: *Robotics and Computer-Integrated Manufacturing* 30.4 (2014), pp. 389–398.

[Por20]   PORSCHE, A. *Dr. Ing. h.c. F. Porsche AG presents innovative 3D-printing technology for bucket seats*. Mar. 17, 2020. URL: https://newsroom.porsche.com/en/2020/products/porsche-3d-printed-bodyform-full-bucket-seat-concept-study-19996.html (visited on 03/23/2024).

[Pot+05]   POTTMANN, H., LEOPOLDSEDER, S., HOFER, M., STEINER, T., and WANG, W. "Industrial geometry: recent advances and applications in CAD". In: *Computer-Aided Design* 37.7 (2005), pp. 751–766.

[PPT09]   PASEK, Z. J., PAWLEWSKI, P., and TRUJILLO, J. "Modeling of Customer Behavior in a Mass-Customized Market". In: *International Symposium on Distributed Computing and Artificial Intelligence 2008 (DCAI 2008)*. Ed. by CORCHADO, J. M., RODRíGUEZ, S., LLINAS, J., and MOLINA, J. M. Berlin, Heidelberg: Springer Berlin Heidelberg, 2009, pp. 541–548.

[Pra+18]   PRADEL, P., ZHU, Z., BIBB, R., and MOULTRIE, J. "A framework for mapping design for additive manufacturing knowledge for industrial and product design". In: *Journal of Engineering Design* 29.6 (2018), pp. 291–326.

[Pri11]   PRINZ, A. "Struktur und Ablaufmodell für das parametrische Entwerfen von Fahrzeugkonzepten". PhD thesis. Logos Verlag Berlin: Technische Universität Carolo-Wilhelmina zu Braunschweig, July 2011.

[Pri+13]   PRINZ, A., NEHUIS, F., VIETOR, T., and STECHERT, C. "The effects of regional specific requirements on the development of vehicle concepts". In: *Conference on Future Automotive Technology*. Springer, 2013, pp. 167–190.

[PS16]   PISCHINGER, S. and SEIFFERT, U. *Vieweg Handbuch Kraftfahrzeugtechnik*. ATZ/MTZ-Fachbuch. Springer Fachmedien Wiesbaden, 2016.

[PS21]   PISCHINGER, S. and SEIFFERT, U. *Vieweg Handbuch Kraftfahrzeugtechnik*. ATZ/MTZ-Fachbuch. Springer Fachmedien Wiesbaden, 2021.

[QD16]   QIAN, X. and DEDE, E. M. "Topology optimization of a coupled thermal-fluid system under a tangential thermal gradient constraint". In: *Structural and Multidisciplinary Optimization* 54 (2016), pp. 531–551.

[Ric+18]   RICHTER, T., WATSCHKE, H., SCHUMACHER, F., VIETOR, T., et al. "Exploitation of potentials of additive manufacturing in ideation workshops". In: *DS 89: Proceedings of The Fifth International Conference on Design Creativity (ICDC 2018), University of Bath, Bath, UK*. 2018, pp. 354–361.

[Roe+21]   ROEHLING, J. D., KHAIRALLAH, S. A., SHEN, Y., BAYRAMIAN, A., BOLEY, C. D., RUBENCHIK, A. M., DEMUTH, J., DUANMU, N., and MATTHEWS, M. J. "Physics of large-area pulsed laser powder bed fusion". In: *Additive Manufacturing* 46 (2021), p. 102186.

[Ros07a]   ROSEN, D. W. "Computer-Aided Design for Additive Manufacturing of Cellular Structures". In: *Computer-Aided Design and Applications* 4.5 (2007), pp. 585–594.

[Ros07b]   ROSEN, D. "Design for additive manufacturing: A method to explore unexplored regions of the design space". In: *18th Solid Freeform Fabrication Symposium, SFF 2007*. Jan. 2007, pp. 402–415.

[RS12]   RIEG, F. and STEINHILPER, R. *Handbuch Konstruktion*. Hanser, 2012.

[RT03]   RIEMER, K. and TOTZ, C. "The many faces of personalization". In: *The customer centric enterprise*. Springer, 2003, pp. 35–50.

[San15]   SANDHANA, L. *Singapore's first 3D-printed urban electric car and tilting three-wheeler ready to race*. New Atlas. Feb. 5, 2015. URL: https://newatlas.com/nv8-nv9-3d-printed-electric-car/35920/ (visited on 03/03/2024).

[Sch13]   SCHÜTZ, T. *Hucho—Aerodynamik des Automobils: Strömungsmechanik, Wärmetechnik, Fahrdynamik, Komfort*. ATZ/MTZ-Fachbuch. Springer Fachmedien Wiesbaden, 2013.

[Sch+17]   SCHUMACHER, A., VIETOR, T., FIEBIG, S., BLETZINGER, K., and MAUTE, K. "Advances in Structural and Multidisciplinary Optimization: Proceedings of the 12th World Congress of Structural and Multidisciplinary Optimization (WCSMO12)". In: (2017).

[Sch+19]   SCHUMACHER, F., WATSCHKE, H., KUSCHMITZ, S., and VIETOR, T. "Goal Oriented Provision of Design Principles for Additive Manufacturing to Support Conceptual Design". In: *International Conference on Engineering Design 2019*. Vol. 1. July 2019, pp. 749–758.

[Sch+20]   SCHMITT, M., MICHATZ, M., FREY, A., LUTTER-GÜNTHER, M., SCHLICK, G., and REINHART, G. "Methodical software-supported, multi-target optimization and redesign of a gear wheel for additive manufacturing". In: *Procedia CIRP* 88 (2020), pp. 417–422.

[Sco16]   SCOTT, C. *Virtual Reality, Artificial Intelligence and 3D Printing Team Up for the Production of the World's First AI-Engineered Car*. July 2016. URL: https://3dprint.com/143821/hack-rod-ai-engineered-car/ (visited on 07/07/2017).

[SD12]   SALEHI- DOUZLOO, V. "An integrated approach to parametric associative design for powertrain components on the automotive industry". PhD thesis. University of Bath, 2012.

[See+12]   SEEPERSAD, C. C., GOVETT, T., KIM, K., LUNDIN, M., and PINERO, D. "A Designer's Guide for Dimensioning and Tolerancing SLS Parts". In: *Solid Freeform Fabrication Symposium*. Austin, TX, 2012, pp. 921–931.

[Sie+16]   SIECOBAN, R., MARTIŞ, R., MARTIŞ, C., HUSAR, C., COADĂLATĂ, L., and Irimia, C. "Multiphysics design, analysis and optimisation platform of PMSM for automotive applications". In: *2016 International Conference and Exposition on Electrical and Power Engineering (EPE)*. 2016, pp. 251–255.

[Sin16]   SINUS, I. *Sinus Milieus Deutschland*. 2016. URL: http://www.sinus-institut.de/sinus-loesungen/sinus-milieus-deutschland/ (visited on 06/22/2016).

[SK15]   SELLE, N. and KLEEMANN, S. "Manuskript zur Vorlesung: Einführung in die Karosserieentwicklung". Verlesungsunterlagen, Sommersemester 2015. 2015.

[SK97]   SPUR, G. and KRAUSE, F. *Das virtuelle Produkt: Management der CAD-Technik*. Hanser, 1997.

[SMT11]   STRICKER, K., MATTHIES, G., and TSANG, R. "Vom Automobilbauer zum Mobilitätsdienstleister". In: *Bain & Company* (2011).

[Spe14]   SPEARS T. *Local motors 3D printed car made full-scale and functional at ITMS 2014*. Designboom. Sep. 12, 2014. URL: https://www.designboom.com/technology/local-motors-3d-printed-car-09-12-2014/ (visited on 03/03/2024).

[SS08]   SCHINDLER, V. and SIEVERS, I. *Forschung für das Auto von morgen: Aus Tradition entsteht Zukunft*. Springer Berlin Heidelberg, 2008.

[Sta15]   STACKPOLE, B. *EDAG's Light Cocoon Is A Metamorphosis for Car Design*. June 18, 2015. URL: https://www.digitalengineering247.com/article/edags-light-cocoon-is-a-metamorphosis-for-car-design (visited on 03/04/2024).

[Sto+20]   STOLFA, J., STOLFA, S., BAIO, C., MADALENO, U., DOLEJSI, P., BRUGNOLI, F., and MESSNARZ, R. "DRIVES–EU blueprint project for the automotive sector—a literature review of drivers of change in automotive industry". In: *Journal of Software: Evolution and Process* 32.3 (2020).

[Str18]   STRATASYS. *The First Fuel-Efficient 3D Printed Car is Back on the Map*. Nov. 13, 2018. URL: https://www.stratasys.com/en/stratasysdirect/resources/articles/3d-printed-car-fuel-efficient-fdm-urbee-2/ (visited on 03/03/2024).

[Str22]   STRATASYS. *Daihatsu Motor Company paves the way for customizing and supplying parts to customers with 3D printing*. Jan. 22, 2022. URL: https://www.stratasys.com/de/resources/case-studies/daihatsu/ (visited on 03/04/2024).

[TEB20]   THORBORG, J, ESSER, P, and BAYAT, M. "Thermomechanical modeling of additively manufactured structural parts—different approaches on the macroscale". In: *IOP Conference Series: Materials Science and Engineering*. Vol. 861. IOP Publishing, 2020, p. 012008.

[TFV21]   TSCHORN, J. A., FUCHS, D., and VIETOR, T. "Potential impact of additive manufacturing and topology optimization inspired lightweight design on vehicle track performance". In: *International Journal on Interactive Design and Manufacturing (IJIDeM)* 15.4 (2021), pp. 499–508.

[Tho09]   THOMAS, D. "The Development of Design Rules for Selective Laser Melting". PhD thesis. University of Wales, 2009.

[Tho+16]   THOMPSON, M. K., MORONI, G., VANEKER, T., FADEL, G., CAMPBELL, R. I., GIBSON, I., BERNARD, A., SCHULZ, J., GRAF, P., AHUJA, B., and MARTINA, F. "Design for Additive Manufacturing: Trends, opportunities, considerations, and constraints". In: {CIRP} *Annals—Manufacturing Technology* 65.2 (2016), pp. 737–760.

[Tof+18]   TOFAIL, S. A., KOUMOULOS, E. P., BANDYOPADHYAY, A., BOSE, S., O'DONOGHUE, L., and CHARITIDIS, C. "Additive manufacturing: scientific and technological challenges, market uptake and opportunities". In: *Materials Today* 21.1 (2018), pp. 22–37.

[Tom+09]   TOMIYAMA, T., GU, P., JIN, Y., LUTTERS, D., KIND, C., and KIMURA, F. "Design methodologies: Industrial and educational applications". In: *CIRP Annals* 58.2 (2009), pp. 543–565.

[Toy15]   TOYOTA. *Tokyo Open Road Project—Toyota i-Road 004*. May 7, 2015. URL: https://pressroom.toyota.com/image/tokyo-open-road-project-toyota-i-road-004/ (visited on 03/04/2024).

[Tra21]   TRAUTMANN, L. "Product Customization and Generative Design". In: *Multidiszciplináris tudományok* (2021).

[Tuc+08]   TUCK, C. J., HAGUE, R. J. M., RUFFO, M., RANSLEY, M., and ADAMS, P. "Rapid manufacturing facilitated customization". In: *International Journal of Computer Integrated Manufacturing* 21.3 (2008), pp. 245–258.

[Tut+13]   TUTUIANU, M., MAROTTA, A., STEVEN, H., ERICSSON, E., HANIU, T., ICHIKAWA, N., and ISHII, H. *Development of a Worldwide harmonized Light duty driving Test Cycle (WLTC)*. Tech. rep. United Nations Economic Commission for Europe (UN-ECE), 2013.

[TWV20]   TSCHORN, J. A., WETZEL, C., and VIETOR, T. "Strategies for Optimising Lap Time and Handling Attributes as Objective Functions in Vehicle Concept Development". In: *41st International Vienna Motor Symposium 22–24 April 2020*. Ed. by VERLAG, V. Vol. 2. 12 813. Bernhard Geringer and Hans Peter Lenz, 2020, pp. 442–458.

[Vaf+21]   VAFADAR, A., GUZZOMI, F., RASSAU, A., and HAYWARD, K. "Advances in Metal Additive Manufacturing: A Review of Common Processes, Industrial Applications, and Current Challenges". In: *Applied Sciences* 11.3 (2021).

[Vaj+18]   VAJNA, S., WEBER, C., ZEMAN, K., HEHENBERGER, P., GERHARD, D., and WARTZACK, S. *CAx für Ingenieure—Eine praxisbezogene Einführung*. 3rd ed. Berlin, Heidelberg: Springer Vieweg, 2018.

[VDI04]   VDI. *VDI-Richtlinie: VDI 2223 Methodisches Entwerfen technischer Produkte*. Deutsch. Tech. rep. ICS: 03.100.40. VDI-Gesellschaft Produkt- und Prozessgestaltung, Jan. 2004.

[VDI14]   VDI. *VDI-Guideline: VDI 3405 Additive manufacturing processes, rapid manufacturing—Basics, definitions, processes*. Tech. rep. VDI-Gesellschaft Produktion und Logistik, 2014.

[VDI18]   VDI. *VDI-Guideline: VDI 3405 Blatt 3.5 Additive manufacturing processes, rapid manufacturing—Design rules for part production using electron beam melting*. Tech. rep. VDI-Gesellschaft Produktion und Logistik, 2018.

[VDI19a]   VDI. *VDI-Guideline: VDI 3405 Blatt 3.2 Additive manufacturing processes— Design rules—Test artefacts and test features for limiting geometric elements*. Tech. rep. VDI-Gesellschaft Produktion und Logistik, 2019.

[VDI19b]   VDI. *VDI-Richtlinie: VDI 2221 Blatt 1 Entwicklung technischer Produkte und Systeme—Modell der Produktentwicklung*. Deutsch. Tech. rep. 2221. VDI-Gesellschaft Produkt- und Prozessgestaltung, 2019.

[VDI19c]   VDI. *VDI-Richtlinie: VDI 2221 Blatt 2 Entwicklung technischer Produkte und Systeme—Gestaltung individueller Produktentwicklungsprozesse*. Deutsch. Tech. rep. 2221. VDI-Gesellschaft Produkt- und Prozessgestaltung, June 2019.

[VDI21a]   VDI. *VDI-Guideline: VDI 3405 Blatt 3.4 Additive manufacturing processes— Design rules for part production using material extrusion processes*. Tech. rep. VDI-Gesellschaft Produktion und Logistik, 2021.

[VDI21b]   VDI. *VDI-Guidelines: VDI 2206 Development of mechatronic and cyberphysical system*. Tech. rep. VDI/VDE-Fachbereich Autonome Systeme & Mechatronik, 2021.

[VDI82]   VDI. *VDI-Richtlinie: VDI 2222 Blatt 2 Erstellung und Anwendung von Konstruktionskatalogen*. Tech. rep. VDI-Gesellschaft Konstruktion und Entwicklung, 1982.

[VDI87]   VDI. *VDI-Guidelines: VDI 2221 Systematic Approach to the Design of Technical Systems and Products—out-of-date -*. Deutsch. Tech. rep. 2221. VDI-Gesellschaft Entwicklung Konstruktion Vertrieb, 1987.

[VDI93]   VDI. *VDI-Richtlinie: VDI 2221 Methodik zum Entwickeln und Konstruieren technischer Systeme und Produkte—zurückgezogen -*. Deutsch. Tech. rep. 2221. VDI-Gesellschaft Entwicklung Konstruktion Vertrieb, 1993.

[VDI97]   VDI. *VDI-Richtlinie: VDI 2222 Blatt 1 Konstruktionsmethodik—Methodisches Entwickeln von Lösungsprinzipien*. Deutsch. Tech. rep. ICS: 03.100.40. VDI-Gesellschaft Produkt- und Prozessgestaltung, June 1997.

[Ves+05] VESANEN, J. et al. "What is personalization?: a literature review and framework". In: (2005).

[VWZ14]   VELEA, M. N., WENNHAGE, P., and ZENKERT, D. "Multi-objective optimisation of vehicle bodies made of FRP sandwich structures". In: *Composite Structures* 111 (2014), pp. 75–84.

[Wag+16] WAGNER, A.-S., KILINCSOY, Ü., REITMEIR, M., and VINK, P. "Functional customization: Value creation by individual storage elements in the car interior". In: *Work* 54.4 (Sept. 2016), pp. 873–885.

[Wat+19] WATSCHKE, H., KUSCHMITZ, S., HEUBACH, J., LEHNE, G., and VIETOR, T. "A methodical approach to support conceptual design for multi-material additive manufacturing". In: *Proceedings of the Design Society: International Conference on Engineering Design*. Vol. 1. 1. Cambridge University Press. 2019, pp. 659–668.

[Web09]   WEBER, J. *Automotive Development Processes: Processes for Successful Customer Oriented Vehicle Development*. Springer Berlin Heidelberg, 2009.

[Web21]   WEBSTER, M. *Term Definition "Methodology"*. online dictionary. May 2021.

[Wei15]   WEISS, C. *World's first 3D-printed supercar aimed at shaking up the auto industry*. July 2015. URL: http://newatlas.com/divergent-microfactories-blade-first-3d-printed-supercar/38201/ (visited on 03/05/2024).

[Wei16]   WEISS, C. *Divergent 3D slices forward with automotive 3D printing*. Nov. 24, 2016. URL: https://newatlas.com/divergent-3d-printing-vehicles/46458/ (visited on 03/05/2024).

[Wer+04] WERNER DANKWORT, C., WEIDLICH, R., GUENTHER, B., and BLAUROCK, J. E. "Engineers' CAx education – it's not only CAD". In: *Computer-Aided Design* 36.14 (2004). CAD Education, pp. 1439–1450.

[WFO09]   WALLENTOWITZ, H., FREIALDENHOVEN, A., and OLSCHEWSKI, I. *Strategien in der Automobilindustrie: Technologietrends und Marktentwicklungen*. ATZ/MTZ-Fachbuch. Vieweg+Teubner Verlag, 2009.

[Wib21]   WIBERG, A. "Design Automation for Additive Manufacturing : A Multi-Disciplinary Optimization Approach". PhD thesis. Linköping University, Machine Design, 2021, p. 62.

[WTH20]   WIESE, M., THIEDE, S., and HERRMANN, C. "Rapid manufacturing of automotive polymer series parts: A systematic review of processes, materials and challenges". In: *Additive Manufacturing* 36 (2020), p. 101582.

[WW12]   WEGNER, A. and WITT, G. "Konstruktionsregeln für das Laser-Sintern". In: *Zeitschrift Kunststofftechnik* 8.3 (2012), pp. 252–277.

[XI18]   XUE, D. and IMANIYAN, D. "A framework for optimal design of complex products". In: *Procedia CIRP* 70 (2018). 28th CIRP Design Conference 2018, 23–25 May 2018, Nantes, France, pp. 416–421.

[YPZ18]   YANG, S., PAGE, T., and ZHAO, Y. F. "Understanding the Role of Additive Manufacturing Knowledge in Stimulating Design Innovation for Novice Designers". In: *Journal of Mechanical Design* 141.2 (Dec. 2018). 021703.

[YZ15]   YANG, S. and ZHAO, Y. F. "Additive manufacturing-enabled design theory and methodology: a critical review". In: *The International Journal of Advanced Manufacturing Technology* 80.1 (2015), pp. 327–342.

[Zhe+17]   ZHENG, P., YU, S., WANG, Y., ZHONG, R. Y., and XU, X. "User-experience Based Product Development for Mass Personalization: A Case Study". In: *Procedia CIRP* 63 (2017). Manufacturing Systems 4.0—Proceedings of the 50th CIRP Conference on Manufacturing Systems, pp. 2–7.

[Zhu+21]   ZHU, J., ZHOU, H., WANG, C., ZHOU, L., YUAN, S., and ZHANG, W. "A review of topology optimization for additive manufacturing: Status and challenges". In: *Chinese Journal of Aeronautics* 34.1 (2021), pp. 91–110.

[Zie12]   ZIEBART, J. *Ein konstruktionsmethodischer Ansatz zur Funktionsintegration*. Bericht: Institut für Konstruktionstechnik. Verlag Dr. Hut, 2012.

[ZW20]   ZHANG, H.-c. and WANG, L. "Application Research of Industrial Product Design Trend in 5G Perspective". In: *E3S Web Conf.* 179 (2020), p. 02042.

MIX
Papier aus verantwortungsvollen Quellen
Paper from responsible sources
FSC® C105338

FSC
www.fsc.org

If you have any concerns about our products,
you can contact us on
ProductSafety@springernature.com

In case Publisher is established outside the EU,
the EU authorized representative is:
Springer Nature Customer Service Center GmbH
Europaplatz 3, 69115 Heidelberg, Germany

Printed by Libri Plureos GmbH
in Hamburg, Germany